Molecular Gastronomy

ARTS AND TRADITIONS *of the* TABLE

ARTS AND TRADITIONS *of the* TABLE:

PERSPECTIVES ON CULINARY HISTORY

Albert Sonnenfeld, *series editor*

Salt: Grain of Life
Pierre Laszlo, translated by Mary Beth Mader

Culture of the Fork
Giovanni Rebora, translated by Albert Sonnenfeld

French Gastronomy: The History and Geography of a Passion
Jean-Robert Pitte, translated by Jody Gladding

Pasta: The Story of a Universal Food
Silvano Serventi and Françoise Sabban, translated by Antony Shugar

Slow Food: The Case for Taste
Carlo Petrini, translated by William McCuaig

Italian Cuisine: A Cultural History
Alberto Capatti and Massimo Montanari, translated by Áine O'Healy

British Food: An Extraordinary Thousand Years of History
Colin Spencer

A Revolution in Eating: How the Quest for Food Shaped America
James E. McWilliams

Sacred Cow, Mad Cow: A History of Food Fears
Madeleine Ferrières, translated by Jody Gladding

Hervé This

Molecular Gastronomy

Exploring the Science of Flavor

translated by M.B. DEBEVOISE

COLUMBIA UNIVERSITY PRESS {*new york*}

COLUMBIA UNIVERSITY PRESS

Publishers Since 1893
New York Chichester, West Sussex

Library of Congress Cataloging-in-Publication Data

This, Hervé.
[Casseroles et éprouvettes. English]
Molecular gastronomy : exploring the science of flavor / Hervé This ; translated by Malcolm DeBevoise.
p. cm. -- (Arts and traditions of the table)
Includes bibliographical references and index.
ISBN 978-0-231-13312-8 (cloth : alk. paper) — ISBN 978-0-231-13313-5 (pbk. : alk. paper)
1. Food--Sensory evaluation. 2. Flavor. 3. Gastronomy.
I. Title. II. Series.

TX546.T5513 2005
664'.072--dc22

20050053784

Columbia University Press books are printed on permanent and durable acid-free paper.

Printed in the United States of America

Designed by Linda Secondari + Vin Dang. *Ferdinand* typeface by Isaac Tobin.

c 20 19 18 17 16 15 14
p 10

It is not enough to know the principles, one needs to know how to *manipulate*.

—*Dictionnaire de Trévoux,* quoted by MICHAEL FARADAY
in the first edition of *Chemical Manipulation* (1827)

Contents

Part Four: A Cuisine for Tomorrow

Series Editor's Preface

"It takes a tough guy to raise a tender chicken!" the late Frank Perdue used to proclaim in his radio and TV advertisements. Physical chemist Hervé This (pronounced teess), the internationally controversial molecular gastronome, explains to us in understandable yet precise terms the science of tenderness.

What defines tenderness, anyway? How does one achieve it in the farmyard and the kitchen? What chemical interactions give a chicken the potential to be a gourmet chicken? How is tenderness perceived by the complex nerve endings and taste buds of the mouth? The current cult of "Slow Food" may have a basis in molecular science, or it may be mere Walden Pondish Romantic Rousseauism. After reading this absorbing book I now know what I mean when I sing, "Try a little tenderness!"

Hervé This combines the seriousness of purpose and acumen of a respected scientist (Collège de France) with the aura of dynamic TV personality. He succeeds more than others in making what seems recondite to some accessible to all. We worry about making good French fries; here we read that there is laboratory predictability in the choice of potato variety, the slicing technique, and the discoloration that occurs when enzymes in the air hit the uncooked spud. Each scientific food inquiry raised in this book takes root in specific everyday (and useful) examples, the whys and wherefores of our very real culinary dilemmas as home cooks and consumers.

Should one salt a steak before, during, or after cooking? We must understand the chemical processes and consequences of that common ritual. How does the shape of the wine glass affect the taste of a given wine? Science gives us real answers. Our molecular gastronome could use equations, but instead he uses words, readably, eloquently, and wittily.

How to cool down a drink that is too hot? Cool your almost-boiling morning Java with cold milk and it will take ten minutes, but wait for the coffee to reach 75°F, then add the milk, and one obtains the same result after only four minutes.

And your *espresso ristretto* (black)? One would think that the energy deflection of the metallic spoon and the diluting effect of sugar would be dramatic. Blowing on the brew proves to be more efficient by half than stirring, even though I had always thought that spreading the heat would lead to more rapid cooling.

The era of culinology, a useful term immediately trademarked in our profit-driven culinary culture by a group called Research Chefs of America, has clearly arrived. How timely, then, is the publication of *Molecular Gastronomy*, the liveliest addition to the growing bibliography exploring culinary science. I might also cite the work of the Monell Chemical Senses Institute and the books of such distinguished scholars as Peter Barham, Harold McGee, and Robert L. Wolke.

Hervé This explores the chemistry, physics, and technology of produce cultivation and selection, food preparation (cooking, freezing), tasting, and digestion in his pioneering TV shows and acclaimed magazine articles. His laboratories at the Institutes for Advanced Research and the seminars he codirects at the Collège de France have attracted many of the celebrity chefs to whose kitchens people flock on pilgrimages of culinary initiation.

This book is as much about the science of eating and enjoying food as about the science of its preparation. How does the brain perceive flavor and decide it is good, and how do we detect textures? Is chewing healthful or even necessary for gastronomic enjoyment? How does breaking down foods by mastication alter their chemistry and release flavors? Study the effect and techniques of flavoring and chewing gum, asserts our media-friendly gastronome, and you'll understand.

One of the most fascinating debates in current food science revolves around the so-called fifth taste, umami. We had been taught that salty, sweet, sour, and bitter are the four taste sensations. As so often happens, new insights come from Asia.

Hervé This traces the search for glutamate receptors, which led to a molecule that university researchers called a metabotropic glutamate receptor. That sounds intimidating! But as I read, This's narrative style took me out of the specialized language of a laboratory I would never dare to enter and into the excitement of scientific understanding of the molecular activity and interaction of the tongue's taste buds and the nose's sensors.

Molecular gastronomy is not a media-driven gimmick; This calls whatever pretentiousness resides therein the necessary price of precision.

Historically, molecular gastronomy is the consequence of the linkage of gastronomy to science in the title and content of Jean-Anthelme Brillat-Savarin's *Physiology of Taste* (1825), made available to us in the splendid translation by M. F. K. Fisher. The science of food, which Brillat-Savarin called gastronomy, was initiated earlier by chemists in the Age of Enlightenment, the late seventeenth and eighteenth centuries, and belongs to the history of science. The kitchen was a laboratory like any other for famous doctor and pioneering chemist Antoine-Laurent Lavoisier. In Germany, Justus von Liebig, working in the Age of Positivism, applied meat extracts to the soups that still bear his name. The test tubes were pots and pans.

Brillat-Savarin called himself "le professeur." Defining gastronomy as the intelligent knowledge of whatever concerns nourishment, the gourmet professor initiates his readers into a veritable eighteenth-century encyclopedia of natural history, physics, chemistry, cookery, business, and political economy.

Hervé This, our new millennium initiator, is more rigorously focused: Molecular gastronomy deals with culinary transformations and the sensory phenomena associated with eating. As a guide he achieves exemplary clarity for the nonscientist reader, and he is consistently entertaining.

ALBERT SONNENFELD

Introduction to the English-Language Edition

THE TITLE I GAVE THE ORIGINAL edition of this book was *Casseroles et éprouvettes:* saucepans and test tubes. Not the sorts of thing one normally expects to find together, either in the kitchen or the laboratory—or so it seemed before the creation of a new scientific discipline called molecular gastronomy. I should perhaps say a word or two about the origin of the name.

In 1988 the Oxford physicist Nicholas Kurti and I were preparing the first of a series of international workshops on the physical and chemical aspects of cooking, and we realized we needed a pithy phrase that would describe this new field of research. Brillat-Savarin's classic definition of gastronomy in the *Physiology of Taste* (1825) naturally came to mind:

> Gastronomy is the intelligent knowledge of whatever concerns man's nourishment.
>
> Its purpose is to watch over his conservation by suggesting the best possible sustenance for him.
>
> It arrives at this goal by directing, according to certain principles, all men who hunt, supply, or prepare whatever can be made into food... .
>
> Gastronomy is a part of:
>
> Natural history, by its classification of alimentary substances;
>
> Physics, because of the examination of the composition and quality of these substances;
>
> Chemistry, by the various analyses and catalyses to which it subjects them;

Cookery, because of the art of adapting dishes and making them pleasant to the taste;

Business, by the seeking out of methods of buying as cheaply as possible what is needed, and of selling most advantageously what can be produced for sale;

Finally, political economy, because of the sources of revenue which gastronomy creates and the means of exchange which it establishes between nations.

(MEDITATION 3, §18)

In this view, the humble hard-boiled egg belongs every bit as much to gastronomy as a wonderfully complicated dish such as the one Brillat-Savarin invented in honor of his mother, *Oreiller de la Belle Aurore,* a kind of pillow made of puff pastry and stuffed with seven kinds of wild game as well as foie gras and truffles. In both cases "intelligent knowledge"—that is, rational or analytical understanding—is needed, but it must be admitted that understanding the scientific principles of boiling an egg will be more useful to many more people. If all you have to eat is an egg, you had better know how to cook it properly.

So we had one part of the name we were looking for. But what kind of gastronomy? The term "molecular" was very fashionable at the time (molecular biology, molecular embryology, and so on), but it was also indispensable if we were to limit the scope of our enterprise. I proposed that we call our field simply molecular gastronomy, but Nicholas thought that "molecular" would too narrowly identify it with chemistry and suggested "molecular and physical gastronomy" instead. We started out calling it by this name, but after a while it seemed too cumbersome. Because the analysis of the structure and behavior of molecules obviously involves a certain amount of physics, after Nicholas's death in 1998 I decided to revert to the shorter form in announcing our workshops, which now meet every two or three years in Sicily. And so molecular gastronomy it has been ever since.

But why not "molecular cooking"? Because cooking is a craft, an art—not a science. Nor is molecular gastronomy the same thing as the technology of cooking, because science is not technology. Furthermore, gastronomy seeks to answer a wider range of questions. For example, why does a tannic wine have a disagreeable taste if one drinks it in the company of salad that has been tossed with an acidic dressing? This question has nothing to do with cooking and everything to do with gastronomy.

What exactly does molecular gastronomy deal with? And how does it differ from the well-established field of food science? Some historical perspective will be useful in answering these questions, but generally speaking it is correct to say that food science deals with the composition and structure of food, and molecular gastronomy deals with culinary transformations and the sensory phenomena associated with eating.

Let's begin by going back to ancient Egypt. When the anonymous author of the London Papyrus used a scale to determine whether fermented meat was lighter than fresh meat, was he doing an early form of molecular gastronomy or of food science? It depends on what the motivation for the experiment was. If he wanted to understand an effect of cooking, it was molecular gastronomy. If he was interested mainly in the properties of meat, then it was food science.

The succeeding centuries witnessed the development of chemistry. For a long time it resembled cooking and used many of the same techniques: cutting, grinding, heating, macerating, and so on. In the late fifteenth century Bartholemeo Sacchi, author (under the pen name of Platina) of a cookbook titled *De honesta voluptate et valetudine* (1475), made little if any distinction between chemistry, medicine, and cooking. More than 250 years later one finds much the same state of affairs in *La suite des dons de Comus* (1742) by François Marin. Note, however, that in the interval the French physician and inventor Denis Papin (1647–1712) had built a pressure cooker in order to recover the substance of bones in broth, hence the name of his machine: the digestor.

Marin's views echoed those of Sacchi. "The science of cooking," he wrote, "involves decomposing, digesting, and extracting the quintessence of meats, drawing from them their light and nutritive juices. Indeed this kind of chemical analysis is the main object of our art." Chemical analysis! We need first of all to make a clear distinction between art, technology, and science. To ride a bicycle, for example, one has to push the pedals forward; this is a matter of technique, or art. If someone were to inquire into the difference between pedaling with the front of the foot instead of the heel, this would be a question of technology (which, as the Greek word indicates, is the systematic treatment of *techne*—art, craft, or skill). And if someone were to survey the surrounding landscape while pedaling a bicycle—say, in order to avoid having to climb a mountain—this would be an example of scientific investigation (or, more generally, the attempt to discover the mechanisms of natural phenomena).

Plainly, then, cooking is not the same thing as molecular gastronomy, for craft aims at the production of goods, not of knowledge. For the same reason molecular gastronomy is no substitute for cooking because it seeks to produce something entirely different. Marin was wrong, then, in saying that cooking is a form of chemical analysis. Similarly, Papin's digestor was an achievement of technology rather than of science.

Let's return to the eighteenth century. In 1773 French chemist Antoine Baumé (1728–1804) devised a "recipe" for preparing "dry stocks for times of war, or stock tablets." In order to improve the extraction of organic compounds he recommended boiling meats again, after the first extraction, then clarifying the stock with egg whites and allowing the liquid to evaporate in a bain-marie until only "perfectly dry and brittle" tablets remained. The whole problem of making stocks and meat extracts was a very important one at a time when neither refrigerators nor freezers existed to preserve food. But there were also financial considerations. It is often forgotten that the great Lavoisier himself took an interest in the confection of stocks. As *fermier général*, responsible for collecting taxes and supplying the hospitals of Paris with food, he understood that it was not the water in a broth that provided nourishment but the matter that had been extracted from the meat and that had undergone chemical reaction in the course of cooking. He therefore devised a densitometer to determine how much meat was necessary to feed the indigent patients of the hospitals.

Three years later, Benjamin Thompson (1753–1814), later Count Rumford (later still he married Lavoisier's widow), published a 400-page book titled *On the Construction of Kitchen Fireplaces and Kitchen Utensils Together with Remarks and Observations Relating to the Various Processes of Cookery and Proposals for Improving That Most Useful Art* (1776). Rumford did a great deal of work related to food and perhaps should be considered one of the most important figures in the prehistory of molecular gastronomy, as he was interested not only in technology but also in science. It is even said that he discovered fluid convection while eating a thick soup whose viscosity prevented the inner layer from cooling, thus causing him to burn his mouth.

Also during this period Antoine-Augustin Parmentier (1737–1813), a pharmacist who had an interest in food, sought to win acceptance for the potato in France and made a study of the flours used to make bread. Parmentier's fame came to be widespread in his native land, which probably explains why A. Viard wrote in *Le cuisinier impérial, ou l'art de faire la cuisine et la pâtisserie pour toutes les fortunes* (1806), "All the arts and sciences have made enormous advances in the last one hundred years, especially chemistry, which has progressed so much that a student familiar with [recent] discoveries about flavor can demonstrate theories whose existence was not previously suspected. It is natural, in these conditions, that cooking, which is a kind of chemistry, advanced at the same pace." Again one sees the confusion between chemistry, a science, and cooking, an art or craft. The definition of artists as inspired craftsmen (proposed by Walter Gropius, founder of the Bauhaus school of design in Weimar Germany) is worth keeping in mind.

Another German, Justus von Liebig (1803–1873), who devoted much of his later scientific career to the analysis of food, made a fortune from an eponymous American company that produced meat extracts from surplus supplies of meat. The chemical theory underlying the product turned out to be wrong, but Liebig's extract became famous throughout the world.

A more important result was obtained shortly afterward by the French chemist Michel-Eugène Chevreul (1786–1889), who analyzed fats and discovered their chemical structure. Also during this period, in Germany, Emil Fischer (1852–1919) was studying sugars, again with significant consequences for the development of chemistry. Already in 1821 Friedrich Christian Accum (1769–1838) had brought out in London a very interesting book called *Culinary Chemistry: Exhibiting the Scientific Principles of Cookery, with Concise Instructions for Preparing Good and Wholesome Pickles, Vinegar, Conserves, Fruit Jellies, Marmalades, and Various Other Alimentary Substances Employed in Domestic Economy, with Observations on the Chemical Constitution and Nutritive Qualities of Different Kinds of Food*. Here the question arises whether a distinction must be made between chemistry as a science and chemistry as an application of science, or technology. My own view is that the terms *chemistry* and *science* should be reserved for the scientific exploration of chemical phenomena.

We now come to the strange case of Louis-Camille Maillard (1878–1936). On completing studies in medicine and chemistry at the University of Nancy, Maillard wrote his doctoral thesis in the latter field on the reaction of glycerol and sugars with amino acids. These chemical processes, first explained in a publication of 1912, are very important because they impart flavor to grilled meats, bread crust, roasted chocolate and coffee, and many other things. After World War I Maillard took up an appointment as professor of biochemistry and toxicology at the University of Algiers, where he taught until his sudden death in Paris almost twenty years later. Curiously, Maillard was renowned for his work throughout the world but not in France until a few years ago.

No brief survey of the prehistory of molecular gastronomy would be complete without mentioning Édouard de Pomiane (1875–1964), a biologist at the Institut Pasteur in Paris who was well known in the first half of the twentieth century for a series of popular works on what he called *gastrotechnie,* or gastrotechnology, an attempt to rationalize cooking similar to ones also being made in the United States and in some European countries. But these works were full of elementary mistakes, based on insufficient experimental evidence. For example, it was believed at the time that the bowl and whisk used to whip egg whites had to be made of copper and galvanized iron, respectively, to promote the formation of foam. But all this had to do with technology, not science.

Food science thus initially developed in close contact with cooking. But it soon gave way to an interest in feeding people and making more efficient use of ingredients. It is often forgotten that until recently the chief concern of people in most countries, even in the West, was having enough to eat. Gradually scientific research came to concentrate more on foods themselves than on their domestic preparation.

But what about the millions of people who cook every day in advanced industrial countries? We now have access to products that have benefited from advances in food science, but do we know how to cook them? This question has two parts. On one hand, how good are the products we use? On the other, how competent are we as cooks?

First, the question of quality. Like so many others in France today who long for the countryside they left in order to live and work in cities, I am not immune to nostalgia for the good old days. I, too, miss the chickens running freely about the courtyard; the asparagus picked just before the meal, with its delicate milky juice running out from the stalk; the peas shelled just before they are cooked; the strawberries served still warm from the sun—all this is the stuff of literature. But the countryside is also the mud that comes when it rains, the wild rabbits that visit at night to undo the gardener's work during the day, the mice that gnaw at the food he has stored away, the aching back that rewards him for his toil.

By all means, then, let us fill our souls with such nostalgia, for they have need of it. But let us also compare. The same Alsatian wine that thirty years ago produced migraine headaches and kept for only four years has now become a nectar that no longer degrades so rapidly. Mediocre homemade yogurt has been supplanted by commercial brands in various flavors that have a perfectly regular texture. Should we reproach them for having a strawberry flavor too unlike the flavor of strawberries from the orchard? Or should we reproach ourselves instead for wanting to eat strawberries in winter? The same goes for insipid year-round tomatoes: Wait for summer!

Enough of this facile apology for progress. We would do better to accept products for what they are and recognize that the possibility of improving them lies first and foremost in submitting them to the transformations of the culinary art. If we want yogurt to be flavored, we should be prepared do it ourselves. In other words, let's go into the kitchen and start cooking.

This is why I raised the second question, concerning our culinary skills. To answer this we need to ask ourselves how we cook, and we will have to admit that by and large we repeat what we have seen done at home, by our parents or grandparents. When we try out a new dish, one that does not belong to the family culinary repertoire, we have the same feeling Christopher Columbus had setting out to discover the New World. Why do most people find cooking so difficult? Because for most people it is a matter of repetition and habit. In Meditation 7, section 48 of the *Physiology of Taste* (best known in English in the translation by M. F. K. Fisher, from whom I quote once again), Brillat-Savarin gave a more detailed answer:

SERMON

"Maître la Planche," said the Professor, in a tone grave enough to pierce the hardest heart, "everyone who has sat at my table proclaims you as a *soup-cook* of the highest order, which is indeed a fine thing, for soup is of primary concern to any hungry stomach; but I observe with chagrin that so far you are but an *uncertain fryer*.

"Yesterday I heard you moan over that magnificent sole, when you served it to us pale, flabby, and bleached. My friend R. ... threw a disapproving look at you; Monsieur H. R. ... averted his gnomonic nose, and President S. ... deplored the accident as if it were a public calamity.

"This misfortune happened because you have neglected the theory of frying, whose importance you do not recognize. You are somewhat opinionated, and I have had a little trouble in making you understand that the phenomena which occur in your laboratory are nothing more than the execution of the eternal laws of nature, and that certain things which you do inattentively, and only because you have seen others do them, are nonetheless based on the highest and most abstruse scientific principles.

"Listen to me with attention, then, and learn, so that you will have no more reason to blush for your creations."

I. CHEMISTRY

"Liquids which you expose to the action of fire cannot all absorb an equal quantity of heat; nature has made them receptive to it in varying degrees: it is a system whose secret rests with her, and which we call caloric capacity.

"For instance, you could dip your finger with impunity into boiling spirits-of-wine, but you would pull it out as fast as you could from boiling brandy, faster yet if it was water, and a rapid immersion in boiling oil would give you a cruel injury, for oil can become at least three times as hot as plain water.

"It is because of this fact that hot liquids react in differing ways upon the edible bodies which are plunged into them. Food which is treated in water becomes softer, and then dissolves and is reduced to a bouilli; from it comes soup-stock or various essences: whereas food which is treated in oil grows more solid, takes on a more or less deep color, and ends by burning.

"In the first case, the water dissolves and pulls out the inner juices of the food which is plunged into it; in the second, these juices are saved, since the oil cannot dissolve them; and if the food becomes dry, it is only because the continuation of the heat ends in vaporizing their moistness.

"These two methods also have different names, and *frying* is the one for boiling in oil or grease something which is meant to be eaten. I believe that I have already explained that, in the culinary definition, *oil* and *grease* are almost synonymous, grease being nothing more than solid oil, while oil is liquid grease."

II. APPLICATION OF THEORY

"Fried things are highly popular at any celebration: they add a piquant variety to the menu; they are nice to look at, possess all of their original flavor, and can be eaten with the fingers, which is always pleasing to the ladies.

"Frying also furnishes cooks with many ways of hiding what has already been served the day before, and comes to their aid in emergencies; for it takes no longer to fry a four-pound carp than it does to boil an egg.

"The whole secret of good frying comes from the surprise; for such is called the action of the boiling liquid which chars or browns, at the very instant of immersion, the outside surfaces of whatever is being fried.

"By means of this surprise, a kind of glove is formed, which contains the body of food, keeps the grease from penetrating, and concentrates the inner juices, which themselves undergo an interior cooking which gives to the food all the flavor it is capable of producing.

"In order to assure that the surprise will occur, the burning liquid must be hot enough to make its action rapid and instantaneous; but it cannot arrive at this point until it has been exposed for a considerable time to a high and lively fire.

"The following method will always tell you when the fat is at a proper heat: Cut a finger of bread, and dip it into the pot for five or six seconds; if it comes out crisp and browned do your frying immediately, and if not you must add to the fire and make the test again.

"Once the surprise has occurred, moderate the fire, so that the cooking will not be too rapid and the juices which you have imprisoned will undergo, by means of a prolonged heating, the changes which unite them and thus heighten the flavor.

"You have doubtless noticed that the surface of well-fried foods will not melt either salt or sugar, which they still call for according to their different natures. Therefore you must not neglect to reduce these two substances to the finest powder, so that they will be as easy as possible to make adhere to the food, and so that by means of a shaker you can properly season what you have prepared.

"I shall not speak to you of the choice of oils and greases; the various manuals which I have provided for your pantry bookshelf have already shed sufficient light for you on this subject.

"However, do not forget, when you are confronted with one of those trout weighing barely a quarter-pound, the kind which come from murmuring brooks far from our capital, do not forget, I say, to fry it in your very finest olive oil: this simple dish, properly sprinkled with salt and decorated with slices of lemon, is worthy to be served to a Personage!

"In the same way treat smelts, which are so highly prized by the gastronomers. The smelt is the figpecker of the seas: the same tiny size, the same delicate flavor, the same subtle superiority.

"My two prescriptions are founded, again, on the nature of things. Experience has taught us that olive oil must be used only for operations which take very little time or which do not demand great heat, because prolonged boiling of it develops a choking and disagreeable taste which comes from certain particles of olive tissue which it is very difficult to get rid of, and which are easily burned.

"You have charge of my domestic regions, and you were the first to have the glory of producing for an astonished gathering an immense turbot. There was, on that occasion, great rejoicing among the chosen few.

"Get along with you: continue to make everything with the greatest possible care, and never forget that from the instant when my guests have set foot in my house, it is we who are responsible for their well-being."

The dissertation of the Professor (as Brillat-Savarin styled himself) is full of errors from the scientific point of view. First of all, there is a difference between boiling temperature and heat capacity. Boiling temperature is the temperature at which a liquid boils. Oil, which begins to decompose before coming to a boil, has a higher boiling point than that of water (100°C [212°F]), which in turn is higher than that of ethanol, the alcohol found in liquor (78°C [172°F]). Heat capacity is something quite different: the quantity of energy needed to increase the temperature of a compound by 1°C (1.8°F).

Furthermore, it is not true that putting something in boiling water always leads to softening, dissolving, and the formation of a sort of purée or mush. When you put egg white in boiling water, for example, it hardens. And although it is true that collagen—the tissue that holds muscle fibers in meat together—dissolves slowly in boiling water, protein coagulation inside these cells produces a tough material.

Nor is it true that the crust formed during frying prevents oil from penetrating meat and helps to concentrate its juices. The crust is full of cracks, through which vapor bubbles escape during the course of frying (they can be seen with the aid of a microscope or in some cases with the naked eye) and through which oil enters. Furthermore, the notion that the juices of a piece of meat go to the center during cooking rests on a misunderstanding. Because these juices are mostly water and therefore not compressible, the effect of cooking a piece of meat (even by boiling in water) is to force them outward, away from the center. What Brillat-Savarin thought was concentration is actually expansion.

Just the same, there is a certain amount of truth in what the Professor says, some of it useful. This is why it is so important to distinguish what is right from what is wrong—to rationalize the old "chemical art" of cooking. Above all we want to retain the idea that "certain things which you do inattentively, and only because you have seen others do them, are nonetheless based on the highest and most abstruse scientific principles." In other words, culinary phenomena—the phenomena that generate transformations in food—are at bottom nothing more than chemistry and physics. To cook well, at least from the technical point of view (art, as I say, is another story), we have to know both these sciences.

This is precisely why Nicholas Kurti and I sought to promote the notion of molecular gastronomy: Chemistry and physics, judiciously applied, can tell us how to preserve the tenderness of meats, how to master the chemical reactions that give the crust of roasted meat its wonderful flavor, and how to avoid the failures that are commonly encountered in making a variety of sauces, from mayonnaise to hollandaise, béarnaise, ravigote, and many others. Do we dare make the leap? We hesitate because as human beings, which is to say primates, we have a fear of new foods or of foods that are unfamiliar to us. Jonathan Swift famously said, "He was a bold man that first eat an oyster." No matter that each new recipe may be likened to the discovery of a new continent, science is there not only to guide us but also to help us exercise our innate capacities for discovery and invention. The application of eternal laws of nature both informs and stimulates culinary innovation.

Is it enough to read cookbooks? Certainly not. They are generally little more than collections of recipes, which is to say protocols that relegate cooks to the status of mere executors. Moreover, they contain a great many doubtful instructions: Steak ought to be seared, because this causes an impermeable crust to form that retains its juices (a sound practice, as it happens, although the reasoning is wrong); in making stocks, meat should be placed in cold water to begin because this will cause "albumen" to coagulate and so prevent the loss of juices; mayonnaise will break if women try to make it when they are menstruating; egg whites will not stiffen if one changes the direction in which they are whisked; and so on. Clearly a bit of science must be brought to bear if we are profitably to explore our culinary heritage, as I propose to do in the pages that follow.

And a bit of poetry as well? Its absence will be regretted only by those who prefer the catastrophes that have routinely been courted by respecting the traditional method for making soufflés rise, for example. It is a mistake to suppose that in understanding physical phenomena we lose the ability to take pleasure in culinary art. Besides, one finds poetry where one may: Are the names "ionone" (for a molecule with a delicate violet odor in dilute alcohol solution) and "hexanal" (for a molecule that imparts a fresh herb flavor to virgin olive oil) any less beautiful than "cauldron" or "knife"?

Poetry aside, let us consider efficiency. Time-honored maxims, proverbs, old wives' tales, folk beliefs, and culinary rules are millstones round our necks that weigh us down when they are false and wings that carry us aloft when they are true. Hence the importance of molecular gastronomy, whose primary objective is first to make an inventory of such rules and then to select those that have withstood careful analysis. Culinary art has everything to gain by separating the wheat from the chaff of empirical observations.

In the first part of this book, then, we will consider the rules that have long guided the preparation of a variety of familiar dishes: stock, hard-boiled eggs, quiches, quenelles, gnocchi, cheese fondue, roast beef, preserves—almost twenty dishes in all.

But anyone who cooks rationally, relying solely on the laws of physics and chemistry, will soon run up against the limits of these two sciences in the kitchen. Take meringues, for example. You want them to rise? Then place them in a glass vacuum bell jar and pump the air out; the air bubbles dilate and the meringues swell and swell, to the point that you with left with "wind crystals." Nothing to chew on there—a culinary disaster.

This leads us to ask ourselves what we like to eat and why. Further questions immediately arise: Why do we stop eating? How many tastes do we perceive? Is flavor modified by changes in temperature?

Modern physiologists of flavor have studied these questions, carrying out experiments suggested to them by their expertise in a particular field of research. They have thrown valuable light, for example, on mastication, the almost unconscious act that, to the civilized mind, separates gluttons from gastronomes. Brillat-Savarin thought this distinction so important that he devoted the first part of the introduction of his book to it. Immediately after the celebrated aphorisms of the preamble, he reports (or possibly only imagines) a "Dialogue Between the Author and His Friend":

FRIEND: This morning my wife and I decided, at breakfast, that you really ought to have your GASTRONOMICAL MEDITATIONS published, and as soon as possible.

AUTHOR: *What woman wants, God wants.* There, in five words, you have the whole guide to Parisian life! But I myself am not a Parisian, and anyway as a bachelor ...

FRIEND: Good Lord, bachelors are as much victims of the rule as the rest of us, and sometimes to our great disadvantage! But in this case even celibacy can't save you: my wife is convinced that she has the right to dictate to you about the book, since it was at her country house that you wrote the first pages of it.

AUTHOR: You know, my dear Doctor, my deference for the ladies. More than once you've complimented me on my submission to their orders. You were even among those who once said that I would make an excellent husband! Nonetheless, I refuse to publish my book.

FRIEND: And why?

AUTHOR: Because, since I am committed to a life of serious professional studies, I am afraid that people who might know the book only by its title would think that I wrote nothing but fiddle-faddle.

FRIEND: Pure panic! Aren't thirty-six years of continuous public service enough to have established the opposite reputation? Anyway, my wife and I believe that everyone will want to read you.

AUTHOR: Really?

FRIEND: Learned men will read you to learn more from you, and to fill out for themselves what you have only sketched.

AUTHOR: That might well be ...

FRIEND: The ladies will read you because they will see very plainly that ...

AUTHOR: My dear friend, I am old! I've acquired wisdom, at least: *miserere mei!*

FRIEND: Gourmands will read you because you do justice to them, because at long last you give them the place they merit in society.

AUTHOR: This one time you're right! It is incredible that they have been misunderstood for so long, the poor fellows! I suffer for them like their own father ... they are so charming, and have such twinkling little eyes!

FRIEND: Moreover, have you not often told us that our libraries definitely lack a book like yours?

AUTHOR: I've said so.... I admit that, and would choke myself rather than take it back!

FRIEND: Now you are talking like a man completely convinced! Come along home with me and ...

AUTHOR: Not at all! If an author's life has its little pleasures, it also has plenty of stings in it. I'll leave all that to my heirs.

FRIEND: But you disinherit your friends then ... your acquaintances, your contemporaries. Have you enough courage for that?

AUTHOR: Heirs! Heirs! I've heard it said that ghosts are deeply flattered by the compliments of the living. That is a divine blessing which I'll gladly reserve for the next world!

FRIEND: But are you quite sure that these compliments will reach the right ghost? Are you equally sure of the trustworthiness of your heirs?

AUTHOR: I haven't any reason to believe that they will neglect one such duty, since for it I shall excuse them from a great many others!

FRIEND: But will they, can they, give to your book that fatherly love, those paternal attentions without which a published work seems always a little awkward on its first appearance?

AUTHOR: My manuscript will be corrected, neatly copied, polished in every way. There will be nothing more to do but print it.

FRIEND: And the chances of fate? Alas, similar plans have caused the loss of plenty of priceless works! Among them, for instance, there was that of the famous Lecat, on the state of the soul during sleep ... his life work ...

AUTHOR: That was, undoubtedly, a great loss. I am far from aspiring to any such regrets.

FRIEND: Believe me, heirs will have plenty to cope, what with the church, the law courts, the doctors themselves! Even if they do not lack willingness, they'll have little time for the various worries that precede, accompany, and follow the publication of a book, no matter how long or short it may be.

AUTHOR: But my title! My subject! And my mocking friends!

FRIEND: The single word *gastronomy* makes everyone prick up his ears. The subject is always fashionable. And mockers like to eat, as well as the rest. And there's something else: can you ignore the fact that the most solemn personages have occasionally produced light works? There is President Montesquieu, for instance!

AUTHOR: By Jove, that's so! He wrote THE TEMPLE OF GNIDUS ... and one might do well to remember that there is more real point in meditating on what is at once necessary, pleasant, and a daily occupation, than in what was said and done more than two thousand years ago by a couple of little brats in the woods of Greece, one chasing, the other pretending to flee ...

FRIEND: Then you give up, finally?

AUTHOR: Me? I should say not! I simply showed myself as an author for a minute. And that reminds me of a high-comedy scene from an English play, which really amused me. I think it's in a thing called THE NATURAL DAUGHTER. See what you think of it.

The play is about Quakers. You know that members of this sect thee-and-thou everyone, dress very simply, frown on war, never preach sermons, act with deliberation, and above all never let themselves be angry.

Well, the hero is a handsome young Quaker, who comes on the scene in a severe brown suit, a big, flat-brimmed hat, uncurled hair ... none of which prevents him from being normally amorous!

A stupid lout, finding himself the Quaker's rival in love, and emboldened by this ascetic exterior and the nature it apparently hides, teases and taunts and ridicules him, so that the young hero grows increasingly furious and finally gives the fool a good beating.

Once having done it, though, he suddenly reassumes his Quakerish manners. He falls back, and cries out in his shame, "Alas! I believe that the flesh has triumphed over the spirit!"

I feel the same way. After a reaction which is certainly pardonable, I go back to my first opinion.

FRIEND: It simply can't be done. You admit that you have shown yourself as an author for a second or two. I've got you now, and I'm taking you to the publisher's. I'll even tell you that more than one friend has already guessed your secret.

AUTHOR: Don't leave yourself open! I'll talk about you in return ... and who knows what I may say?

FRIEND: What could you possibly say about me? Don't get the idea that you can scare me off!

AUTHOR: What I shan't say is that our native land prides itself on having produced you; that at twenty-four you had already published a textbook which has since become a classic; that your deserved reputation inspires great confidence in you; that your general appearance reassures the sick; that your dexterity astounds them; that your sympathy comforts them. All this is common knowledge. But I shall reveal to the whole of Paris (*here I draw myself up*), to all of France (*I swell with oratorical rage*), to the Universe itself, your only fault!

FRIEND (*gravely*): And what is that, may I ask?

AUTHOR: An habitual vice, which all my exhortations have not corrected.

FRIEND (*horrified*): Tell me! Don't torture me like this!

AUTHOR: You eat too fast!

[HERE THE FRIEND PICKS UP HIS HAT, AND EXITS SMILING, FAIRLY WELL CONVINCED THAT HE HAD MADE A CONVERT.]

The latest studies of the present generation of physiologists of flavor are reported in the second part of the book. These studies, which, unlike Brillat-Savarin's literary investigations, constitute the true physiology of flavor, have a direct usefulness in the kitchen for those who are bold enough to apply them. Yet a more complete science remains to be built on this base. Recall the definition I gave earlier of molecular gastronomy. Only the dictums, proverbs, and rules that have been shown to be sound can be placed at the heart of the new science that is needed. This means that culinary practice must henceforth be based on a genuine physiology of flavor. In asking how are individual dishes to be prepared, however, we are dealing with something rather different: the modeling of culinary operations. And this modeling must be based on an understanding of the physical transformations to which foods are subjected in cooking.

This is the subject of the third part of the book. However, note that the "intelligent knowledge" the reader will find there must be judged with reference to the peculiar situation in which cooks find themselves. *Sutor ne supra crepidam iudicaret:* Let not the cobbler criticize [a work of art] above the shoe. But cooks have no choice but to judge above and beyond the pan, for they know that this work is not for the stomach but for the heart and the soul. And this is why seemingly useless explorations find their place here—for love of the beauty of pure knowledge.

Consider the egg yolk—the ordinary egg yolk, which generations of cooks have used without looking at it any longer than was necessary to prevent it from being broken or spilled. Yet it possesses a complex and unsuspected structure, which science has disclosed to us. No longer can the sight of an egg yolk be thought uninteresting. For boredom is born not of uniformity but of a certain offhandedness, a lack of deference. Thanks to science, which teaches us that even the yolk of an egg deserves to be the object of curiosity and admiration, we have no reason to be bored in the kitchen again.

Obviously molecular gastronomy does not aim solely at attaining pure knowledge of this sort, because it seeks also to give practical knowledge a sound basis by explaining why successful recipes work and why mistakes occur. Thus, for example, if you ask why lumps form when flour is placed in a hot liquid, you will at once be led to useful conclusions that allow certain culinary techniques to be rationalized and refined.

I propose a new article of faith: Whoever understands the reasons for the results he or she obtains in the kitchen can improve on them. This is why I devote so much attention at the outset to criticizing traditional recipes. What if one follows a recipe for making mayonnaise to the letter, but the sauce breaks? As a hostage to the recipe one will have no alternative but to throw out the offending egg, mustard, and oil. But the cook who understands that mayonnaise is an emulsion—a dispersion of oil droplets in water (from the yolk and vinegar)—will be able to save the sauce, not by adding another egg (as the leading authorities advise) but by decanting the oil and once again dispersing it in the watery ingredients.

The whole third part of the book therefore is concerned with improving recipes and preparations. Reasoned analysis, allied with the ideal of perfectibility, is what gives cooking its soul. The spirit of Brillat-Savarin lives on.

In exploring physical and chemical mechanisms of cooking we will find ample grounds for modifying classic recipes. Consider once again the soufflé. First we test the maxim that the whipped egg whites must be firm; then we analyze the way in which we perceive the light, airy texture of the dish. From this we draw conclusions about what a soufflé ideally ought to be: Rational analysis of the classic recipe shows that a soufflé should be heated from below. But now we find ourselves confronted with an awkward situation. Recent studies—experimental and theoretical alike, because the two obviously go together—have shown that the classic method of cooking soufflés in the oven is not indispensable. What are we to do? Do we abandon the oven, out of distrust of tradition, in order to produce a better soufflé? Or do we go on following the teachings of the old masters, forgetting that things such as mayonnaise and puff pastry, centuries ago, were themselves innovations?

The fourth part of this book frankly rejects conservatism and resistance to change in the name of another tradition: intelligent knowledge. It is in the name of this tradition that we undertake to devise new chocolate mousses, that we resolve to abandon the useless clarification of stocks, that we generalize from the traditional aioli (a garlic emulsion) to produce a new class of flavored mayonnaises, and that, despite opposition from defenders of tradition who fear the temptations of novelty, we dare to conduct chemistry experiments in our own kitchens.

The fact of the matter is that we do both chemistry and physics whenever we make an emulsified sauce or grill a piece of meat. Nonetheless, we are like Molière's Monsieur Jourdain, not realizing that we have been doing chemistry and physics all along. What is more, satisfied with what we have achieved, we do not look for ways to achieve something better. In the fourth part of the book it is therefore the soul of cooking that I insist on. For in seeking to understand the reasons for what we do in the kitchen we seek not to poison ourselves but rather to enjoy flavors that until now we have only dreamed of. Let us go about our cooking, then, with full knowledge of what it actually involves.

I Secrets of the Kitchen

PART ONE

I

Making Stock

Meat loses its juices no less readily when it is placed first in boiling water than in cold water.

BEEF STOCK (ALSO CALLED BROTH OR BOUILLON), chef Jules Gouffé wrote in *Le livre de cuisine* (1867), is "the soul of ordinary cooking." Rather than water, which has no taste, cooks have long used wine or the liquid obtained by simmering meats and vegetables in water. This liquid traditionally is served as a first course, but it is used also to moisten various dishes and as an element in the preparation of sauces. How should it be made?

Cookbooks are filled with admonitions. "The gradual heating of the liquid," states *Le livre de cuisine de Mme. E. Saint-Ange* (1927), "is of the highest importance for the clarity as well as for the flavorfulness of the broth." The idea that meat ought to be cooked in water that is initially cold was advanced almost a century earlier by Marie-Antoine Carême, perhaps the most famous of all French cooks, also known as the "cook of emperors" for his service in Russia to the Tsar and in England to the Prince of Wales. Carême proposed an explanation in *L'Art de la cuisine française au XIXe siècle* (1833): "The broth must come to a boil very slowly, otherwise the albumin coagulates, hardens; the water, not having had the necessary time to penetrate the meat, prevents the gelatinous part of the osmazome from detaching itself from it."

Some ten years before Carême, Brillat-Savarin wrote that "to have a good broth, it is necessary that the water heat slowly, in order that the albumin does not coagulate inside [the meat] before being extracted; and it is necessary that the boiling be scarcely perceptible, so that the various parts that are successively dissolved are able to blend intimately and readily." The release of the meat's juices is related, then, to the clarity of the broth.

Are we justified in supposing that the meat yields different quantities of juice depending on whether it is initially placed in hot water or cold water? Plainly everything depends on the length of cooking. Gouffé argued that the cooking must last several hours: "There comes a moment when the meat is cooked and has nothing more to give you in the way of juice or aroma. To let it remain in the pot after it has been completely exhausted by cooking, far from improving the broth, risks spoiling it. I advise a limit of five hours for a large *pot-au-feu*."

It is hard to understand why, after five hours of cooking, the smell and taste should still depend on the temperature of the water at the outset. On the other hand, it is easy to see that the mechanical agitation of the broth places in suspension particles that have been separated from the meat, leading to a cloudy broth that will then have to be clarified, at the risk of weakening its flavor.

Since 1995 my colleagues and I have been comparing various cooking methods for broths. It soon became apparent that broths begun with boiling water were cloudier. Nonetheless, the problem of the initial temperature of the water persisted, despite the work of Justus von Liebig (1803–1873), universally known for his pioneering studies in organic chemistry and for his broths and meat extracts.

Liebig claimed that the essential nutrients of meat are found not in its muscle fibers but in its fluids, which are lost during roasting and broth making. When the meat is plunged into boiling water and the temperature then reduced to a simmer, Liebig wrote in an article published in 1848, "the albumin immediately coagulates from the surface [of the meat] inwards, and in this state forms a crust or shell, which no longer permits the external water to penetrate into the interior of the mass of flesh. But the temperature is gradually transmitted to the interior, and there effects the conversion of the raw flesh into the state of boiled or roasted meat. The flesh retains its juiciness, and is quite as agreeable to the taste as it can be made by roasting; for the chief part of the sapid constituents of the mass is retained, under these circumstances, in the flesh." Conversely, in order to produce a good broth, he recommended against putting the meat in hot water because otherwise the juices would be confined in the meat and one would end up with a tasteless broth.

These precepts formed the basis for a commercial enterprise. Using a vacuum to evaporate the broth produced by cooking minced meat in cold water, Liebig obtained a "beef extract" that he sold throughout the world, simultaneously propagating the theory that broth is best made starting from cold water. Liebig was a good chemist, but on this subject he did no more than copy the writings of Brillat-Savarin, who was neither a scientist nor a cook, a half-century earlier. There is no question that plunging meat into boiling water blanches it immediately. But is this method less effective in extracting the juices?

Lost Juices

Nothing beats an experiment: Let's cut up a big piece of meat into two equal parts and place half of it in cold water and the other half in boiling water; then let's heat and periodically weigh the two pieces. We will find that the mass is very rapidly reduced in the boiling water and more slowly in the cold water.

After about an hour of cooking, however, the two pieces have lost the same amount (give or take a gram). After this point the mass no longer varies, even with several additional hours of cooking. Moreover, in a blind tasting the two broths are indistinguishable: Liebig's broth theory, doubtful in principle, turns out to be false in practice.

However, our experiment suggests a new way of treating the boiled meat once its juices have been released. If it is left to cool sufficiently in the broth, its mass can increase by more than 10% because the meat absorbs the liquid. Why not let the meat cool in a juice made from truffles, for example?

2

Clarifying Stock

Does convection suffice to clarify a broth?

COOKBOOKS MAKE MANY CURIOUS CLAIMS. One by the late Bernard Loiseau asserts that adding ice cubes to a cloudy broth "stuns" the suspended particles, causing them to fall to the floor of the stock pot. One may quibble with this way of stating the matter, but does it contain an element of truth?

The Modeling of Broth

Let's begin by selecting particles close in size to those that actually cloud beef broth. Ground coffee is a good candidate because it consists of particles of various sizes. But because, unadulterated, it would excessively tinge the color of the broth, let's dilute it by running water through the grounds until the coloring agents are rinsed out. The result is a black powder of mixed granularity.

Let's now divide this powder into two equal parts and put them in identical glasses containing the same quantity of water. After we heat the contents of the two glasses in a microwave oven, the particles suspended in both liquids reveal the presence of energetic currents that cease after a few seconds. Now, very carefully, put ice cubes in one glass and in the other a mass of hot water equal to that of the ice cubes. Nothing happens in the latter case, but in the glass with the ice cubes the suspended powder shows signs of intense agitation.

The observed motion is not surprising: The ice cubes cool the water in the upper part of the glass while melting and releasing cold—that is, dense—

water. This dense water falls, and the hot water at the bottom rises and cools on contact with the ice cubes, which are warmed in turn, and so on until the ice cubes have melted.

What happens to the particles? Have they all been "floored" by a knockout blow from the cold water? Not really. A strange segregation appears: Although both the large and small particles in suspension are carried downward by the convective current and seem to be deposited at the bottom of the glass, in reality the upward current carries the smallest particles back with it. Why don't the large particles rise again as well? Probably because particles that fall in a fluid have a maximum speed.

If the liquid were immobile, the particles would be subject to two forces: the weight of the particles themselves, pulling them toward the bottom, and the upward thrust—or buoyancy—described by Archimedes' principle (equal to the weight of the fluid displaced by the particles). Ultimately the particles wind up forming a deposit because they are denser than the water and because the resultant of these forces pushes them toward the bottom.

Yet the falling particles are subject to another force that slows them down. The intensity of this drag depends on the viscosity of the liquid and the radius and velocity of the particles. To simplify matters, let's begin by considering this force in isolation, acting independently of the others. During free fall, the force is initially zero (because the rate of fall is zero), and the particles are accelerated by the downward force. Gradually, however, the upward drag asserts itself, offsetting the resultant of the weight and the Archimedean thrust so that the particles end up falling at a constant speed, which is their maximum velocity.

The Segregation of Particles

When the liquid rises, after having fallen to the bottom of the glass, it tends to carry both the small and the large particles with it. However, these particles have different maximum rates of fall and therefore react differently. Because their radius is small, the small particles fall to the bottom at a speed that is less than that of the fluid's upward motion. By contrast, the large particles, with their greater maximum speed, fall too fast for the ascending fluid to be able to carry them back up again; they remain at the bottom of the glass.

How can we test this hypothesis? Would it be possible to reproduce the experiment in a more viscous fluid and in this way modify the maximum velocity of the particles? Marc Fermigier of the École Supérieure de Physique et de Chimie in Paris, citing Darcy's law for flow through porous media, observes that the increase in viscosity may slow the speed of convection and thus alter the phenomena being studied: In pure water the current pushes the particles back up along the curve of its path, below the convection cell, because it is able to penetrate the granular medium of these particles; the more viscous the fluid, however, the harder it is for the current to pass between them.

Fermigier's colleague Eduardo Weisfred adds that the large particles, being more inert than the small ones, tend to shift to another line of current below the convection cell, so that they are carried toward areas where the current is less swift and where sedimentation is possible; the small particles, on the other hand, follow the fluid throughout its complete motion.

What is the practical lesson of all this? That only the large particles form a significant deposit, and the small ones will continue to cloud the bouillon—just as the lamb, the innocent polluter in La Fontaine's fable, is said to have fouled the wolf's drinking water.

3

Hard-Boiled Eggs

How to center a yolk and get the cooking time just right.

COOKBOOKS SAY THAT TO OBTAIN a hard-boiled egg with a centered yolk, the egg must be cooked in water that has already been brought to a boil. Experience often demonstrates the soundness of this advice, but sometimes one follows it and the yolk ends up being off center. Other times the yolk comes out in the center when the egg has initially been placed in cold water. What good is advice if it isn't always good?

First things first. Why is the yolk sometimes decentered? Because it has changed position inside the egg. Why has it changed position? Because it is subject to the forces of gravity and buoyancy. Do these forces push the yolk upward or downward? It is often supposed that the yolk is more dense than the egg white. Let's find out whether this is true by doing an experiment. In a tall, slender glass—a transparent "shell" that will permit us to see the respective positions of the yolk and the white—place a yolk and, on top of this, four or five whites. The yolk slowly rises, which is easily enough explained by the fact that it contains lipids, or fats, which are less dense than the water that makes up most of the egg white.

Does the fact that the yolk comes to float atop the whites account for its decentering in the hard-boiled egg? We can test this hypothesis by another experiment. First, place an egg on its side in a saucepan filled with water. Then bring the water to a boil and let the egg cook for ten minutes or so, making sure that it remains motionless throughout. On peeling off the shell of the hard-boiled egg we find that the yolk has moved toward the top. Let's repeat the experiment with an egg that has been left in a vertical position long enough for the yolk to rise. If we cook the egg in the same position we discover that the yolk is decentered once again toward the top. These experiments confirm that the difference in density between the yolk and the white is responsible for the position observed.

"Yes, but what about the membranes that center the yellow in the egg?" those who know of their existence will ask. The experiment with the glass removes these membranes and so eliminates their effect, it is true. But the cooking experiment demonstrates that they are not sufficient to hold the yolk in place.

How, then, can we reliably obtain a cooked egg with the yolk in the center? The answer may be deduced from the preceding experiments: We must prevent the yolk from rising in the shell. How? By closing off the vertical axis of movement, and with it the yolk's ability to float. In practice this means manipulating the egg while it cooks. Put one in boiling water, rolling it around in the saucepan for about ten minutes, and then remove the shell. You will find that the yolk is centered.

The same experiment, only starting with cold water, produces a centered yolk as long as one rolls the egg around for a longer time, which becomes tedious after a while. The cookbooks are partially correct, then, but they offer no insight into the forces actually at work. The key to success, it turns out, lies in not allowing the yolk to stay still.

Hamine Eggs

What is the recipe for a perfectly cooked egg? The question may seem odd because tastes vary so greatly. Some people like the dark green ring on the surface of the yolk, for example, and others detest the sulfurous odor that accompanies it. Cookbooks say to cook the egg for only ten minutes in boiling water, without any further explanation. Let's look first at the problem of cooking time.

Why ten minutes rather than, say, five or fifteen? Because after five minutes the egg has not yet hardened, and after fifteen minutes it is assumed that the egg white will be rubbery and the yolk sandy. Yet the latter result is not universally observed throughout the world. Hamine eggs prepared in

Jewish communities in Greece and elsewhere are famous for their tenderness, although they are cooked for several hours. How do these cooks avoid the sulfurous smell of overcooked eggs? And why does the yolk in their eggs not take on the greenish color found in eggs cooked longer than ten minutes in France?

These questions give rise to others. How do eggs cook in the first place? The white consists of about 10% proteins (amino acid chains folded upon themselves in the shape of a ball) and 90% water. During cooking the proteins partially unfold (they are said to be "denatured") and bind with each other, forming a lattice that traps water—in other words, a gel.

The tenderness of the cooked egg white depends on the quantity of water trapped (the loss of a part of this water is what makes overcooked fried eggs rubbery and overcooked yolks sandy) and on the number of proteins making up its lattices (more lattices mean that more water is trapped, rigidifying the entire system).

Experiments in which eggs are cooked at different temperatures provide a clue to the mystery of hamine eggs. When an egg is cooked in boiling water, at a temperature of 100°C (212°F), not only does its mass progressively diminish as water is eliminated from the gel that forms, but many kinds of protein coagulate as well. By contrast, when an egg is cooked at a temperature just a bit higher than the temperature at which its proteins have all coagulated—about 68°C (154°F)—it thickens (thanks to the coagulation of only a few proteins) while retaining its water, a guarantee of tenderness and smoothness.

Hamine eggs traditionally are cooked in embers, with a temperature range of only 50–90°C (122–194°F). There is therefore no paradox, only a good empirical understanding of the coagulation of the egg's proteins during cooking. Cooking eggs in boiling water nonetheless has an advantage: Because the temperature is constant, one obtains a constant result by fixing the cooking time. However, this temperature does not take into account the nature of the egg.

Isn't it time that we avail ourselves of the benefits of modern technology? High-quality thermometers now make it possible to cook eggs at temperatures closer to those at which their proteins are denatured: at 62°C (144°F) one of the proteins in the white (ovotransferrin) is cooked, but the yolk remains liquid because the proteins that coagulate first in this part of the egg require a temperature of 68°C (154°F). Obviously this would mean longer cooking times, but the result is a perfectly cooked egg.

4

Quiches, Quenelles, and Puff Pastries

Their expansion is caused only indirectly by the eggs they contain; cooking vaporizes the water, which puffs up the dough.

A QUICHE MUST NOT BE COOKED TOO LONG, or it will lose its smooth consistency. It will not be perfectly moist (or *chevelotte,* as they say in some parts of Lorraine), hence the culinary rule: Cooking must stop when the quiche begins to rise. Rise? Why does it rise? And why is this a sign that the quiche is done?

To answer the question, let us examine another dish that also expands: the quenelle, a cousin of the German *Knödel,* or dumpling. Whether the quenelle is made of finely minced fish or meat, the flesh has in most cases been mixed with cream, eggs, and panada (in this case a dough obtained by kneading flour with boiling water). Once poached the quenelles are placed in a sauce and, if all goes well, they puff up, even when the eggs have not been beaten.

The quenelle and the quiche therefore have cream and eggs in common. Which of the two ingredients is necessary to make them rise? Heating the cream does not appear to make it expand. But a glance at recipes for *petits choux,* composed of moist bread and eggs, reveals that the egg is the common denominator of dishes that puff up. Some chefs knew this a century ago: The anonymous author of a cooking manual published in 1905 wrote that "dishes that contain eggs can puff up."

Why should eggs have this property of expansion? A moment's reflection yields an insight that even long experience making quiches, varying the proportions of egg white, yolk, cream, and bacon, may not: Although there is no reason why air bubbles should spontaneously appear in the heated egg, eggs do contain water (almost 90% in the case of the white, 50% in the yolk), and this water evaporates during cooking. Anyone who has ever fried an egg has seen the white part rise on contact with the hot skillet. Steam becomes trapped under the coagulated layer and pushes it up.

Perfectly Puffed Up

Now that we know the reason for the egg's expansion, how can we maximize it? If one were to broil an egg white, cooking it from above, the water in the top part would evaporate and escape without puffing it up. By contrast, if one heats the white from below, as in the case of a fried egg, one observes the expansion I have just described. Dishes that are supposed to puff up therefore must be heated from below. Soufflés? Put the ramekin on the floor of the oven. Macaroons? Quenelles? Use a bottom-heated metal plate instead of a broiler pan.

How much can a dish that contains egg swell up through the vaporization of water alone? An egg white, which weighs about 30 grams and therefore contains about 27 grams of water (not quite an ounce), can generate more than 30 liters of steam (almost 32 quarts). Why, then, should such expansion be limited to soufflés, macaroons, *pains à l'anise* (sometimes called jumbles or knots in English-speaking countries), cheese-filled puff pastries, and the like? Because a significant part of the moisture is lost by diffusion through the upper part of the dough; indeed, careful scrutiny of these items as they cook reveals the escape of vapor bubbles. Maximum expansion would require an impermeable upper layer that keeps all the moisture inside.

Return to Quiche

Let's come back to quiche. Why is its expansion a sign that it is done? The liquid that runs around the bacon and into the pastry is composed mainly of fat, water, and proteins. The proteins coagulate as a result of cooking, binding together to form a gel, a lattice that traps the water and fat. The more water there is, the softer the gel. (By contrast, meat, fish, and eggs whose water has entirely evaporated are tough solids.) In other words, although it does puff up a quiche, the evaporation of water is also a signal that the quiche is beginning to lose its tenderness. So there is some truth to the old saying.

But is it the whole truth? No, for the puffing up occurs only if one cooks the quiche by heating it from below. And so the maxim must be revised to say that a quiche is done when, having been heated from below, it begins to puff up. Then, and only then, will it be *chevelotte.*

5
Échaudés and Gnocchi

Is it true that when they float to the surface of the cooking water they are done?

ÉCHAUDÉS ARE VERY OLD PREPARATIONS: As early as 1651 Nicolas de Bonnefons mentions small pieces of dough that have been "scalded" in boiling water. There are many recipes, but from the oldest échaudés to potato gnocchi and *gnocchi à la parisienne* the principle is the same: One begins with a dough composed of starch, egg, and water. In the case of potato gnocchi, extra starch is contributed by the granules found in the potatoes (which are cooked, peeled, and mashed). Other recipes include parmesan cheese and milk. The dough is kneaded with a spatula, then placed on a floured baking sheet and rolled into cylinders, which are cut up into small sections and pressed against a fork to give them their classic shape. Next comes the cooking. The échaudés or gnocchi are put into a saucepan containing salted water or broth, and when they rise to the surface they are said to be cooked. Is this maxim trustworthy? If so, why? The question is not trivial, for many preparations call for the same type of ingredients and the same type of cooking: Alsatian *spätzle,* various central European dumplings, and so on.

If we put some échaudés or gnocchi in boiling water we discover that they fall to the bottom of the saucepan. Then they gradually swell and become lighter. At first they follow the convection currents in the saucepan, continually coming back to the bottom. But after a while—thirty seconds or so—they begin to float. It is therefore accurate to say that échaudés and gnocchi rise to the surface.

Are they actually cooked at this point, as the old saying has it? If you taste them, you will find that they are edible. But are they really done?

Denser, on Average

Let us analyze the problem for classic gnocchi, which are made of potato, flour, and egg. This dough preserves its form in the course of cooking because the egg coagulates on contact with the boiling water, the temperature being higher than 68°c (154°f). The potato has been cooked beforehand, with the result that the starch granules that fill its cells have swelled up. As for the flour, it is composed of starch granules and gluten; the latter is made of proteins, which with the kneading of the dough form a network that embeds the starch granules in the cellular structure of the potato. The starch contained in flour, like that of the potato, is insoluble when cold, but in hot water the suddenly porous starch granules absorb the water molecules.

This mechanism explains the swelling of gnocchi during cooking. And because the density of the starch is greater than that of water (when flour is mixed with water it falls to the bottom of the container), the total density of the gnocchi diminishes, approaching—without ever quite reaching—that of the water. Why, then, do the gnocchi wind up floating on the surface? It must be that some substance whose density is lower than that of water, either air or steam, is incorporated with it in the form of bubbles. What sort of experiment would determine which one it is?

If we cook some gnocchi in water from which the dissolved air has been expelled by heating the water for a long time (the bubbles that form at the bottom of the pan in the initial stage of heating and then rise to the surface of the water are air bubbles, not steam bubbles), we find that the gnocchi nonetheless rise. So it is not air that makes them do this.

The only other possibility is that water vapor forms bubbles that cling to the gnocchi and cause it to rise. Remove one of the gnocchi that have come to the surface of the degassed water, gently roll it on a cutting board in order to puncture any steam bubbles, and then put it back into the water, and it sinks again. Once its surface crevices have again been filled with microscopic steam bubbles, it will rise to the top once more.

If you want to see this with your own eyes, cook some cauliflower florets. Their surface is so irregular that it efficiently traps the steam; a glistening gaseous layer can be seen to cover the florets just when they begin to float. Pat the surface and you will release the steam, causing the florets once again to descend to the bottom.

A Qualified Maxim

Our investigation of échaudés and gnocchi is not yet finished, for we still do not know whether the fact that they float means that they are cooked. When *is* an échaudé cooked? When the egg has coagulated? When the starch has stiffened? To answer this question, let's form échaudés of various sizes—half a centimeter and 10 centimeters in diameter, for example. Let's then cook them together and measure the internal temperature when they rise to the surface of the water. We find that the temperature of the small échaudé is much higher than that of the large échaudé. This proves that buoyancy is not a reliable sign that échaudés are cooked or, at least, if one judges according to this criterion, that échaudés of different sizes will be cooked to different degrees.

Moreover, the temperature of the largest échaudé sometimes turns out to be lower than that at which starch stiffens and egg coagulates. In other words, the simple fact of buoyancy is insufficient. If one aspires to a standard of precision worthy of the great chefs, it is necessary to prolong cooking beyond the point at which they come back up or to draw up a table that gives cooking temperatures as a function of an échaudé's size.

6

The Well-Leavened Soufflé

Water evaporates upon contact with the heated sides of the ramekin and causes the soufflé to rise.

HOW CAN ONE MAKE A PERFECT SOUFFLÉ every time? The question is not only of practical interest to cooks. Measuring the pressure inside a cheese soufflé and the loss of water during cooking will also help food scientists understand the dynamics involved.

Contrary to a widely held belief, this dish is within the reach of beginners. In the case of a cheese soufflé, for example, begin by preparing a béchamel sauce, heating flour and butter, and then adding to this milk and grated cheese. Next, off heat, fold in egg yolks and (very carefully) beaten egg whites. The result is cooked in the oven at a temperature of 180–200°C (about 375–425°F), depending on the size, for twenty to thirty minutes.

It has long been supposed that soufflés rise because they contain air bubbles that expand upon heating. A simple calculation shows that this effect can generate an expansion of only about 20%. However, expert pastry cooks are perfectly capable of making soufflés that double or triple in volume.

Soufflés rise because of the vaporization of the water found in the milk and eggs. The proof is readily seen: When one cuts into a soufflé with a knife in order to share it with a party of delighted dinner guests, a cloud of steam bursts forth. At exactly this moment, to everyone's great dismay, the soufflé collapses.

A Model Soufflé and a Soufflé Model

What is the maximum possible expansion of a soufflé? A simple thermodynamic model is useful in thinking about this question. The warm air inside the oven transfers its heat first to the ramekin and to the upper part of the soufflé, establishing a temperature gradient between the periphery and the center of the mixture. When the temperature reaches 100°C (212°F) near the sides of the ramekin and at the soufflé's upper surface, the water inside evaporates and a crust is formed.

This account is supported by internal temperature readings. When one sinks the probe of a thermocouple (a very precise sort of thermometer) into a soufflé during cooking, one finds that the upper crust has the same temperature as the oven and that at the center the temperature falls before rising again near the bottom.

Moreover, this measurement reveals a hidden dynamic at work in soufflés. If the probe is placed at a fixed position in relation to the ramekin, it registers first an increase in temperature, then a slight decrease or leveling off, and finally a jump back up to about 70°C (158°F). What is happening is that as the heat gradually becomes distributed throughout the soufflé, the evaporation of the water as it comes into contact with the bottom of the ramekin produces bubbles that push the whole of the soufflé upward, which in turn causes cold layers of the mixture to rise to the level of the probe. These layers nonetheless continue to heat up, and the cooking is done when the eggs have thoroughly coagulated, at about 70°C.

In Search of New Heights

How does the water evaporate? Comparison of two soufflés that are identical except for the egg whites used in each suggests an answer: A soufflé with very stiff egg whites rises higher than one whose whites have been whipped for a shorter time because steam bubbles have a harder time penetrating the stable foam of the vigorously beaten whites. Thus leavening is caused by at least two things: the formation of steam bubbles and the trapping of this moisture in the body of the soufflé, in areas where the temperature is sufficiently great to avoid recondensation.

How much water is needed to increase the volume of a soufflé to its greatest extent? Weighing a soufflé before and after cooking reveals that about 10% of its mass has been lost, which is to say that for a soufflé that weighs 300 grams (about 10 ounces), 30 grams (about 1 ounce) can evaporate. Keep in mind that 1 gram of liquid produces 1 liter, or slightly more than a quart, of vapor. Nonetheless, not all the water in a soufflé remains trapped inside; otherwise the internal pressure would exceed a hundred atmospheres. Recent measurements have shown that the pressure increases during cooking by only a few dozen millimeters of mercury, which proves that only part of the evaporated water is retained; the rest escapes in the form of bubbles that eventually burst at the surface of the soufflé.

This suggests that the way to obtain a perfectly leavened soufflé is to heat the bottom of the ramekin, to use very firm whipped egg whites, and to seal the surface in order to prevent the release of the bubbles formed inside. How would one go about doing this? One possibility would be to place the soufflé under a broiler before putting it in the oven. This method has the additional advantage that the soufflé then rises in a regular fashion and, when it is done, has a smooth golden glaze on top that promises a rich flavor.

7

Quenelles and Their Cousins

They're best cooked slowly after the dough has been chilled and allowed to rest.

AS WITH ÉCHAUDÉS, often called gnocchi today, there are many recipes for fish quenelles, but whether they call for salmon or trout or pike they are all variations on a theme: To the finely ground flesh of the fish one adds fat (beef kidneys, butter, or cream) and perhaps egg and panada (either bread soaked in milk or a dough made by combining flour with boiling water). The ingredients are kneaded for a long time—so long, in fact, that Isabella Beeton (author of the famous cookbook published in England in 1860 as *Beeton's Book of Household Management*) wrote, "French quenelles are the best in the world, because they swell up more." And they swell up more, she explained, because they are kneaded longer.

Why should kneading quenelles have anything to do with their succulence? And why should quenelles hold their shape during cooking, even when they do not contain any egg? Florence Lefèvre and Benoit Fauconneau at the Institut National de la Recherche Agronomique (INRA) in Rennes have indirectly answered the question by exploring the thermogelling properties of river (or brown) trout.

The fleshy tissue of the trout is composed of cells, or muscle fibers, that contain myofibrillary proteins. These proteins, which are responsible for muscle contraction, form a gel when they are heated in a water solution. Like the proteins in egg whites, the proteins in trout muscle tissue bind together, creating a network that traps water. In a quenelle, this gel also traps fat and the expanded starch granules contributed by the panada.

Understanding the chemistry of gelatinization allows us to make quenelles and various other products from farm-raised salmon. These products, which Norwegian companies hope to bring to market soon, would be culinary cousins to Asian fish noodles and surimis (dumplings made from freshwater fish such as carp, especially in China). In France, where farm-raised trout is more common, the proteins of this fish are being studied with a view to creating new products as well.

Which proteins form these gels? Like all cells, muscle fibers contain sarcoplasmic proteins that regulate cellular function and maintenance. But they also contain specific myofibrillary proteins, of which the main ones are actin and myosin. In water solution, Lefèvre showed, only the myosin gels alone. The actin by itself does not gel, although incorporating it in a myosin preparation was found to increase the rigidity of the gel.

Under what conditions does gelatinization take place? In the case of quenelles, as in other dishes that depend for their effect on myofibrillary protein gels, the practical problem is how to combine the greatest possible tenderness with sufficient firmness. The parameters that determine the firmness of a gel are the storage time of the solution, the rate of heating, and the maximum cooking temperature, in addition to the protein concentration, acidity, and salt concentration of the solution.

To study the effect of these factors, the biochemists in Rennes inserted the pointed tip of a penetrometer into the trout with constant pressure and measured the degree of deformation. Having first established that this test gauges firmness in the same way biting into the flesh of a fish does, the researchers went on to analyze the gels formed by heating different protein solutions and discovered that the maximum protein concentration was on the order of 10 grams per liter.

A Well-Deserved Rest

Firmness depends also on the length of time solutions are stored, for it is during this time that protein interactions begin to form a gel. Its firmness changes during cooking. A few minutes' heating within a range of 70–80°C (158–176°F) is enough to stabilize the incipient gel, but prolonged cooking results in a loss of water and therefore of tenderness. A rate of heating of 0.25°c per minute has been found to produce a sufficiently firm and elastic gel for making quenelles.

Because proteins contain ionizable lateral groups, their behavior depends especially on the acidity of the solution in which they are placed: In an acidic environment, the acid groups of the proteins are unchanged, but the base groups bond with a hydrogen ion, positively charging the protein molecules and causing them to repel one another rather than to combine. Conversely, in an insufficiently acidic environment, the base groups are neutralized while the acid groups are ionized, likewise producing a repulsion. Thus the acidity of the solution determines the bonds not only between proteins but also with water molecules. The optimal acidity levels depend on the proteins involved and on the animal species from which these proteins come. The INRA chemists showed that, in the case of river trout, the formation of gels is optimized when the acidity of the protein solution is higher (a pH of about 5.6) than the levels conducive to gelatinization in other fish.

This research makes it possible, finally, to perfect the classic preparation of quenelles. First, the quenelle dough must be chilled and left to rest for a few hours, so that a gel forms from the proteins released by the ground muscle fibers. The quenelles themselves should then be heated gently, in a very low oven. Finally, if the quenelles have been slightly acidified, the firmness this imparts will yield a more tender result through the addition of extra water (which in this case means a strongly flavored liquid such as shellfish fumet or fish stock) during cooking.

8

Fondue

How to choose wines and cheeses so that the fondue never flops.

DOES THE TRUE CHEESE FONDUE come from Savoy in France, or the Valais in Switzerland, or the canton of Fribourg? How many types of cheese should be used? One? Two? Four? Connoisseurs passionately disagree. Wars have been started for less. Physical chemistry may not permanently settle such disputes, but it should at least enable lovers of the dish to reach agreement over why, despite its simplicity, the fondue sometimes flops. Athony Blake, director of food sciences and technologies for the Firmenich Group in Geneva, has discovered a surefire way to prevent it from turning into a solid mass lying at the bottom of the pot beneath a greasy liquid.

A fondue is no more than cheese heated with wine. The combination of water (from the wine) and water-insoluble fat (from the cheese) means that the successful fondue is necessarily an emulsion, a dispersion of microscopic droplets of fat in water solution. The fondue therefore is a cousin to béarnaise and hollandaise sauces, which are also obtained by the fusion and dispersion of a fatty substance (in this case butter) in an aqueous phase or zone (from vinegar and egg yolks).

In a béarnaise sauce, the fat droplets are coated by tensioactive molecules found in the egg yolk, in such a way that the water-soluble (hydrophilic) part of these molecules is exposed to the water and the water-insoluble (hydrophobic) part to the fat. The surface-active molecules that cover the fatty droplets in a fondue are known as casein proteins, which are already present in the milk, itself an emulsion, and which combine to form aggregates called micelles. These aggregates are made up of several types of casein, bound together by calcium (especially phosphate) salts. One of the caseins, the kappa-casein, typically lies outside the micelles and ensures their mutual repulsion (because of the negative electrical charge they bear). This repulsion is important for the stability of the milk, for it prevents the coalescence of the fatty droplets covered by the micelles.

In cheesemaking, the rennet that is added to the milk contains an enzyme that detaches a part of the kappa-casein, triggering the aggregation of micelles into a gel in which the fatty matter is trapped. Cheese therefore seems an unlikely candidate for reviving an emulsion in the fondue, having been formed from a milky emulsion that has deliberately been ruined. It nonetheless lends itself to this purpose because it has been aged and mixed with wine.

Aging and Viscosity

Connoisseurs of fondue know that the success of the dish has to do particularly with proper cheese selection. Questions of flavor come into play as well, but well-ripened cheeses are best suited to the preparation of fondues because, in the course of aging, enzymes called peptidases have broken up the casein and the other proteins into small fragments that are more readily dispersed in the water solution. These casein fragments then emulsify the fatty droplets and increase the viscosity of the aqueous phase (which is why a Camembert fondue, for example, will always turn out well).

This increase in viscosity is analogous to the heretical practice of thickening a fondue by adding flour or any other ingredient containing starch, such as potatoes. Swelling up in the warm aqueous solution, the starch granules increase its viscosity and limit the motion of the fatty droplets, which thus are kept separate from one another. In this way the emulsion—which is to say, the fondue—is stabilized.

To Doctor or Not to Doctor

Connoisseurs challenge this practice on the ground that it changes the taste of the dish, insisting instead on the skillful combination of cheeses and wines. They select very dry wines—indeed, wines that are excessively acidic and, if possible, very fruity. Why are these properties useful? Athony Blake has shown that such wines have high concentrations of tartaric, malic, and citric acids. Malate, tartrate, and especially citrate ions are very good at chelating (or sequestering) calcium ions. The acidic and fruity wines experts prefer help separate the casein micelles and release their constituent proteins, which stabilize the emulsion by coating the fatty droplets.

Chemists have devised ways to tweak the classic recipe for fondue, for example by adding bicarbonate of soda, which neutralizes the acids and encourages the formation of calcium-chelating ions. Another option, if one suspects that the wine contains too little tartaric, malic, or citric acid, is to add some; the best choice is citric acid, in its salt form, in a proportion of 1–2%. Do this and you can be sure your fondue will be a success.

9

Roasting Beef

Allowing meat to rest after cooking causes the juices that have been retained in its center to flow outward to the dry periphery.

MANY HOME COOKS TODAY are pressed for time. Their haste prevents them from eating good roasts of beef, for example, for they often neglect an indispensable step after cooking: letting the meat rest, with the door of the oven open. Omitting this step means that the meat will be tough and dry. Professional cooks are well aware that letting meat rest is essential if it is to be tender, but they believe this is because, in the process of cooking, the juices want to "escape" the heat of the oven and consequently flow back into the center of the meat. Letting meat rest afterward therefore is seen as a way of ensuring that its fluids will be thoroughly redistributed. Why should this be true? What really happens when we roast a piece of meat?

Let's examine the matter with the cold and clinical eye of a physical chemist. Chemists know that meat is composed of cells, or muscle fibers: sacs filled mainly with water that contain molecules responsible for metabolism and contraction. These cells are sheathed with collagen and grouped together in bundles, which themselves are grouped together in larger bundles. Naturally this is a simplified description; animal muscle also contains fatty matter, blood, and so on.

How does the muscle structure react when it is heated? Because heat is introduced into a roast by conduction, air being heated in an oven to a temperature of about 200°C (392°F), the water evaporates from the outer layer inward to a point where the temperature is 100°C (212°F). This crusty, desiccated outer layer is thin. Closer to the center, the temperature slowly rises during cooking, and the structure of the meat is transformed by degrees because the various proteins in the meat coagulate at different temperatures. From 70°C (158°F), for example, myoglobin, which transports oxygen in the blood, is oxidized: The ferrous iron it contains is transformed into ferric iron, with the result that the meat turns pink. At 80°C (176°F) the cell walls begin to break down, bringing the myoglobin into contact with oxidant compounds and causing the meat's color to change to brown.

Can blood accumulate in the heart of the roast? When a temperature of 50°C (122°F) is reached in the outside layer, the collagen contracts, compressing the juices inside (although the degree of compressibility is small because the juices are mainly water) and expelling the juices of the periphery outward. The center of the roast, composed of liquids and largely incompressible solids, cannot receive these juices. Anyone who is not convinced of this has only to roast a few pieces of beef, weigh them, and determine their density before and after cooking.

Good Advice, Bad Reasoning

These steps are instructive. First of all, one notices that a roast shrinks when cooked in the usual manner, losing almost a sixth of its weight. This loss results from the elimination of the meat's juices, which are expelled by both contraction of the collagen and evaporation of peripheral water. Note that this observation fatally undermines the theory of cauterization, which holds that the coagulated surface of the meat seals in its internal juices. Near the turn of the twentieth century, for example, Mme. E. H. Gabrielle, author of *La cuisinière modèle,* remarked, "Put the roast on the spit before a very hot fire, in order to sear and tighten the pores of the meat, which thus conserves its juices." Similarly, the great French chef August Escoffier (1846–1935) wrote in his book for home cooks, *Ma cuisine* (1934), that the purpose of browning is "to form around the piece a sort of armature that prevents the internal juices from escaping too soon, which would cause the meat to be boiled rather than braised." Both views are mistaken. Not only does the notion of "pores" have no anatomical basis, but measurement shows that the loss of juices actually increases with cooking.

Empirical analysis also establishes that juices do not flow back to the center of the roast; the density of the cooked center does not differ significantly from that of raw meat. This means that the center has undergone little or no modification during cooking (which is not surprising in the case of French-style roasts, which remain almost raw in the center) and also that the center is full of juices compressed by the shrinking of the collagen.

Why, then, is it a good idea to let meat rest after cooking? Weighing the center and outside portions of roasts, we find that the cooked center loses more juice while resting than do the peripheral parts (which, having already been dried out, are less likely to lose any). Letting the meat rest therefore does redistribute juices from the center outward so that the outer parts regain their tenderness. But this is not because cooking had previously forced fluids to flee to the center.

Given that the juiciness of the meat depends on the amount of juice it has, why not use a syringe to reinject the juices that have drained out from the roast during cooking? Seasoned with salt and pepper, these juices would give the meat a taste it never had—except in the old days, when cooks used to lard meats before cooking with seasoned pieces of bacon.

10

Seasoning Steak

As with the controversy in *GULLIVER'S TRAVELS* over how to crack open an egg, there are two opposing schools in the matter of how to grill a steak: those who salt it before cooking and those who salt it afterwards.

WHEN YOU GRILL A STEAK, naturally you salt it. But when? Before putting the meat on the grill? During cooking? Just before eating it?

Cooks are naturally inclined to respond on the basis of their own experience, but sometimes this is insufficient. As Oscar Wilde remarked, experience is the sum of all our past errors; as long as errors are not recognized, they remain alternative truths. Therefore it helps to conduct experiments in which the various parameters are controlled—the only way to cut to the heart of things, meat among them.

Some argue that introducing salt beforehand gives it time to penetrate, so that the meat is seasoned, if not quite all the way through, then at least much of the way. Others are equally convinced that salting meat before cooking causes its juices to be drawn out by osmosis. Meat is composed of cells—muscle fibers—that contain water, proteins, and all the other molecules necessary to cellular life. If the meat is placed in contact with salt at the outset, the naysayers claim, then the fact that the concentration of water in the meat is greater than in the surrounding layer of salt means that the water will osmotically migrate toward the salt, drying out the meat.

But that is not all. Not only would the meat gradually lose its juices, but during the course of their escape it would be partly boiled in them, instead of being grilled, so that it would not brown properly. It would also lose tenderness, which depends on the concentration of water in food. Some participants in the seminar on molecular gastronomy that I have been conducting in Paris for several years have mentioned a harmful effect on the internal color of the meat, noting that the juice that comes out from the meat is made up largely of blood (along with intracellular water). However, advocates of the water loss theory sometimes forget that muscle fibers are sheathed in a supporting tissue known as collagen.

The structure of different cuts of meat (beef ribs, beefsteak, pork chops, and so on) is so varied that the question must be refined. Let's consider two simple and useful examples, a thin piece of red meat such as steak and the white meat of a fowl, and measure three things: the rate at which the salted meat discharges ("sweats") water, the amount of weight lost, and the residual amount of salt in the meat.

Coating, Sprinkling, and Sweating

Let's begin by considering the first question: How much juice comes out of the meat in the presence of salt? Drench pieces of red and white meat in table salt and let them sit, weighing them at regular intervals. The results probably will vary depending on the meat selected, which may have been cut along the axis of the fibers (with the grain), perpendicularly (against the grain), or diagonally (across the grain). Naturally, water will drain out more readily if the fibers have been opened up. In the case of the red meat, the type of steak matters, too. For example, a rib steak ought to lose more than a flank steak.

With flank steak we find that the discharge of water is very slow, whereas white meat such as chicken loses 1% of its weight in the first thirty minutes after salting. Of course, what happens to salted meats left to sit at room temperature is very different from what occurs during cooking, but the results are plain enough: There is no disgorging of liquid, even though the meat has been coated with salt. In the case of actual cooking, when one would season it with only a small amount of salt, the purging action would be weaker still. Thus it appears that salt has no notable effect—a provisional but nonetheless probable conclusion. You can salt your flank steak when you like, without fear of its drying out.

Turning to the second question, whether the salt penetrates the steak during cooking, consider the experiments I have conducted in collaboration with Rolande Ollitrault of the École Supérieure de Physique et Chimie in Paris and

Marie-Paule Pardo and Éric Trochon of the École Supérieure de Cuisine Française. We salted the same cut of meat before and after cooking, measuring the loss of juice and, most importantly, analyzing the pieces of cooked meat with a scanning electron microscope and a device for detecting chemical elements by means of X-rays.

X-ray analysis reveals the presence of various chemical elements (notably sodium and chlorine in the case of kitchen salt), making it possible to determine whether the salt diffuses through the meat. Again, the answer is clear: Rather than penetrating to the center, it actually passes out of the meat during cooking. On the other hand, when a piece of meat that has been trimmed of fat is placed on the grill, a very small amount of metal is observed to enter the outer layer of the meat.

The nature of meats is so varied that the more subtle effects of preparation and cooking may make themselves felt only insofar as they suit our desires and answer to our illusions. "Nature," in the sybilline words of Leonardo da Vinci (anticipating Hamlet), "is full of infinite reasons that were never in experience." This does not mean that experimentation must be abandoned. It means that experiments must be carefully designed so that the fire of truth may be discovered beneath the smoke of subjective experience and individual opinion.

II

Wine and Marinades

Beef marinates better in red wine than in white wine.

IT IS SAID THAT FISH MUST BE COOKED in white wine but that red wine should be used to marinate and cook tough meats in order to tenderize them. It is also said that parsley must not be used if the marinating process lasts more than two days and that one should not roast marinated meats because roasting dries them out. How far should we credit these familiar dictums?

Japanese physical chemists recently provided partial corroboration. Experiments conducted some twenty years ago in France, at the Institut National de la Recherche Agronomique station in Clermont-Ferrand, showed that beef is tenderized by prolonged immersion in acid solutions, which dissolve collagen and various other proteins principally responsible for the toughness of raw meats while ionizing these proteins, increasing the amount of water they retain. Vinegar is not the sole ingredient of such marinades, however, and the role of wine in particular remained mysterious. Two researchers in the department of home economics at the University of Koshien in Japan, Kazudo Okuda and Ryuzo Ueda, completed the French study after first examining the effect on meats of fermented products such as vinegar and soy sauce.

Preliminary investigations established that the mass, water content (the tenderness of meats depends particularly on their juiciness, which is to say their water content and the ease with which this water is released), and texture of meats that were boiled after having first been marinated in these solutions were modified by five ingredients: alcohol, organic acids, glucose, amino acids, and salt.

Decisive Marinades

In 1995 Okuda and Ueda extended this study by analyzing samples of beef that were boiled after having been marinated in white wine, in red wine, and in solutions containing only certain components of wine. The samples used (cubes of meat weighing about 50 grams, or a bit less than 2 ounces) were marinated for three days, then boiled for ten minutes or so. The outer part and the inner part were analyzed separately. The water content and mass of the samples marinated in red wine were slightly greater than those of the samples marinated in white wine, but the masses of dry matter were about the same. In other words, red wine marinades did a better job of preserving the tenderness of the meats.

Furthermore, the maximal resistance to internal compression was clearly lower in the case of the samples marinated in red wine, which is to say that red wine marinades also did a better job of tenderizing the meats. Finally, after cooking, both the inner and outer parts of the samples marinated in red wine were more tender than the ones marinated in white wine and also more tender than samples cooked without having been marinated.

How are these advantages to be explained? Okuda and Ueda previously showed that the effects of sugars, amino acids, and inorganic salts are weak but that red wines contain more polyphenols than white wines because they are generally more tannic and highly colored (tannins and anthocyanins, the natural color pigments in wine, are polyphenols). Because polyphenols react chemically with proteins, the two researchers tested their effect by marinating the same meats in trial solutions containing fixed concentrations of tannic acid (representing the polyphenols), organic acids, and ethanol.

Solutions composed of water, ethanol, and organic and tannic acids (such as the ones found in red wine) modified the meat in the same way as red wine alone, suggesting that red wine marinades act primarily through organic acids and tannic acid; ethanol and, to a lesser degree, organic acids are important during cooking.

The molecular details of these reactions are being studied, but it appears that proteins react with the polyphenols found in red wines in such a way as to seal the juices of the meat by hardening, or caking, its surface. Classic recipes therefore are justified in recommending that a combination of red wine and acidic liquids, such as vinegar, be used for tenderizing meats. It now remains to elucidate the mysteries associated with the roasting of marinated meats and the proper use of parsley.

12

Color and Freshness

How to prevent discoloration in fruits and vegetables.

THE VIBRANT COLORS OF FRUITS AND VEGETABLES are a sign of their freshness. Alas, no sooner have avocados, salsifies, apples, pears, and mushrooms been sliced or chopped than they turn brown. Can this degradation be avoided? Can fresh-squeezed apple juice make it from the kitchen to the table without turning dark? Cooks have long recommended using lemon, whose juice they believe prevents the appearance of colors associated with overripe, damaged, or rotten organic matter. Is this recommendation sound?

Let's put it to the test. If we compare avocado slices exposed to the oxygen in the air with slices that have first been sprinkled with lemon juice, the difference is plain after a few hours. This confirms the wisdom of customary culinary practice but does not explain why lemon juice has a protective effect. If acidity were responsible, it ought to be possible to substitute vinegar. But this is easily disproven by experiment.

And so? Lemons contain ascorbic acid, or vitamin C, an antioxidant compound. Pure ascorbic acid of the kind one finds in tablet form at the pharmacy ought to be more effective than lemon juice, and experiments show that this is indeed the case. By investigating the role of oxygen in the darkening of vegetables, modern food science has been able to add to the empirical list of remedies that cooks have compiled, which includes not only the juice of certain citrus fruits (lemons, oranges, limes) but also various salty brines.

Vitamin C Versus Enzymes

The darkening of vegetables is caused by enzymes called polyphenol oxidases, which alter the structure of the polyphenol molecules of fruits and vegetables. These molecules have a benzene center surrounded by six carbon atoms at the apices of a hexagon, with either a hydrogen atom or a hydroxyl (–OH) group associated with each carbon atom. In the presence of oxygen the polyphenol oxidase enzymes replace the hydroxyl groups with oxygen atoms, producing quinones whose reaction generates brown pigments of the same family as melanin (the pigment that is formed in our skin when it tans under the sun). This enzymatic darkening, thought to defend plants against ravaging birds and insects, is observed in most fruits, leaves, and many mushrooms that have been cut; it is not commonly found in uncut vegetables because the enzymes and polyphenols are separated by membranes.

Various methods are used to prevent the darkening of vegetables and fruits that have been sliced or chopped—often their lot in the kitchen. Freezing and refrigeration slow but do not prevent the action of enzymes. Pasteurization, a more radical procedure that inactivates the enzymes, cannot be applied to all fruits and vegetables, for it often degrades their texture and color. Finally, vacuum packing—sealing fruits and vegetables in containers from which the oxygen has been drawn out—prevents the appearance of brown compounds; alternatively, nitrogen and carbon dioxide atmospheres sometimes are used in the food processing industry.

Inhibitors are also found in nature that, in minute proportions, work to prevent enzymatic darkening. For example, a very weak dose of salicylhydroxamic acid completely inhibits the formation of polyphenol oxidases in apples and potatoes. Bentonite, a protein-absorbent clay, reduces the activity of enzymes as well. Gelatin, activated charcoal, and polyvinyl pyrrolidone can also be used also extract soluble phenols from wines and beers, but they modify the properties of these beverages.

Sulfites are used in the food processing industry to prevent darkening because they bond with quinones and form colorless sulfoquinones. Sulfur dioxide and sodium metabisulfite are commonly used by wine producers, for example, but they too have secondary effects that worry health authorities (in addition to migraine headaches caused by excessively sulfited wines, asthma attacks and outbreaks of hives, nausea, even anaphylactic shock have been reported). Research into other equally effective but less harmful inhibitors therefore is needed.

Other inhibitors have been discovered or synthesized whose safety remains to be proved. One that is now being studied is cysteine—an amino acid containing sulfur—and its derivatives, as well as natural compounds (found in honey, figs, and pineapple) and synthetic ones.

New Protective Agents

Jacques Nicolas, a researcher at the Conservatoire National des Arts et Métiers in Paris, is exploring the use of cyclodextrins. Working with researchers at the Institut National de la Recherche Agronomique station at Montfavet, Nicholas first analyzed the antidarkening effect of these molecules in trial solutions containing one or two phenols and some polyphenol oxidases from the Red Delicious apple. For the time being, until commercial applications of these results are developed, if you want to make apple juice without resorting to inert atmospheres you will need to clarify the juice. Polyphenol oxidases in the cellular chloroplast of apples form solid fragments that darken, aggregate, and fall to the bottom of the liquid as sediment. So let the juice rest awhile, and then decant the clear amber liquid.

13

Softening Lentils

The virtues of sodium bicarbonate.

CONSIDER THIS PASSAGE from an anonymous work published in 1838 under the title *Le cuisinier parisien:* "Beans, peas, and lentils, and many other vegetables cook well only in very pure and light water; [water] from rivers and streams is always the best; that from wells is worthless. In places where only well water is to be had, it can be made suitable for cooking vegetables by adding to it a little carbonate of soda, dissolved in water to the point that it no longer whitens the water. It leaves a small deposit; one takes the clear liquid and uses it to cook the vegetables." Why should the quality of the water determine the tenderness of vegetables? And why should carbonate of soda (or wood ash, also recommended by the same author) be useful?

Basic Softening

Ash and bicarbonate have in common the property of making water basic, or alkaline. To determine whether this alkalinization acts on vegetables, we may begin by cooking lentils in three identical pans over the same heat. In the first pan let's use distilled water as the cooking medium; in the second, water made basic by the addition of sodium bicarbonate; and in the third, water that has been made acidic by adding a bit of vinegar.

The difference in tenderness between the three samples is so clear that no laboratory instrument is needed to measure it: When the lentils in pure water are just cooked, the ones in acidified water are still as hard as pebbles, whereas the others in water enriched by sodium bicarbonate are falling apart. Thus the relative acidity of the cooking medium determines the rate at which the vegetables are softened.

Why should this be? Vegetables are composed of cells held together by parietal tissue, which is composed of pectin and cellulose. To soften this tissue, one must therefore modify its pectic "cement." In an acid environment, pectin molecules are neutralized: Their $-COO^-$ carboxylate groups capture the hydrogen atoms, yielding electrically neutral $-COOH$ groups. Because the pectin molecules are no longer subject to electrostatic repulsion, the lentils remain hard. By contrast, sodium bicarbonate triggers the ionization of the $-COOH$ carboxylic acid groups into $-COO^-$ groups, with the result that electrostatic repulsions between the pectin molecules cause them to separate, breaking down the parietal walls and thus softening the lentils.

The Hardness of Water

Is the change in the acidity of the water used to cook the lentils the only effect of sodium bicarbonate? *Le cuisinier parisien* indirectly suggests that the hardness of the water, caused by the presence of calcium ions, plays a part as well. To test the effects of these ions, let's once again use two identical pans, adding lentils and putting them over the same heat, then pour distilled water into the first pan and water that has been artificially hardened by the addition of calcium carbonate into the other one. After about 45 minutes the lentils cooked in pure water are done, but the lentils cooked in the hard water are still as hard as wood. This time, the effect results from the calcium ions, whose two positive charges bind them with phytic acid molecules and pectin molecules, reinforcing molecular cohesion rather than weakening it. Monovalent ions, such as sodium, do not establish such bonds.

These phenomena are of particular interest to lentil producers, who are looking for ways to make their products easier to prepare. Can lentils be precooked, for example? At the Institut National de la Recherche Agronomique station in Montfavet, Patrick Varoquaux, Pierre Offant, and Françoise Varoquaux have studied this question with a view to identifying the optimal conditions for softening such seeds without releasing either the starch they contain or, by rupturing the molecular structure of the integument, cellular fragments. Steam cooking would be a good way to achieve this result if it did not take so long.

The Montfavet researchers cooked lentils at different temperatures for different lengths of time and measured the firmness of the seeds. They observed first that the firmness diminished with the length of cooking time, in inverse ratio to temperature. This came as no surprise, but further quantitative analysis proved to be instructive. The curves charting the firmness of the seeds as a function of time, for each temperature, display an elbow shape, which indicates that the softening consists of two phenomena. The first is rapid and probably is associated with the diffusion of the water toward the interior of the lentils; the second is slower and seems to be associated with the gelling of the starch in the hot water, which causes a starchy paste to form.

The researchers also observed that the percentage of lentils that burst open during cooking increases exponentially as a function of time as the temperature rises above 80°C (176°F). Temperature thus affects both the integrity of the lentils and their firmness: At temperatures above 86°C (187°F), the proportion of lentils that fall apart exceeds the proportion of lentils that become soft while retaining their form—hence the culinary rule suggested by these studies, namely that lentils should be cooked at a temperature lower than 80°C (187°F). Of course, this takes time.

14

Souffléed Potatoes

Analysis of a classic dish shows how to avoid the greasiness of deep frying.

SOUFFLÉED POTATOES LOOK LIKE SMALL, crispy golden balloons. They are said to have been discovered on August 25, 1837, during the dedication of the railroad line linking Paris and Saint Germain-en-Laye. The menu for the official luncheon was to include fried slices of potato, but when the train had trouble climbing the last hill the chef was forced to interrupt the frying; once the guests were finally seated, he immersed the slices once again in very hot oil in order to make them crispy. They puffed up.

Since then cooks have differed over the proper way to make this difficult masterpiece of classic French cuisine. Physicochemical analysis has recently illuminated the mechanisms that cause the potatoes to puff up and revealed how to limit the absorption of oil by the potatoes during frying.

Cookbooks do not say why the recipes they give for souffléed potatoes should work. It has long been claimed that this dish and the ideal thickness of the sliced potatoes were studied by the French chemist Michel-Eugène Chevreul (1786–1889), a pioneer in the chemistry of fats. The story is plausible, given the importance of heated fat in this dish, but I have found no trace of any such investigation in the works of Chevreul. Four years after Chevreul's death, however, chef Auguste Colombié noted in his *Éléments culinaires à l'usage des demoiselles* (1893), "Thanks to the good offices of M. Decaux, the gracious and learned laboratory assistant of the late Chevreul, who kindly furnished me with the necessary thermometers, I was able to make three scientific experiments on the puffing up of potatoes, Wednesday 14 April 1884, at the warehouse showroom of the Compagnie Parisienne du Gaz." There follow several pages in which Colombié presents the results of his experiments, with no reference to Chevreul. It therefore seems probable that historians of cookery have identified Colombié with Decaux and Decaux in turn with Chevreul.

The Technology of *Soufflage*

How should souffléed potatoes be prepared? Most traditional recipes recommend cutting the potatoes lengthwise into slices between 3 and 6 millimeters (1/8 and 1/4 inches) thick. The slices are washed, dried, and then cooked in oil that has been heated to a temperature of 80°C (176°F). Once the slices have risen to the surface, after six or seven minutes, they are removed from the oil and allowed to cool before being put back and cooked a second time, only now at a higher temperature. The authors of these recipes attribute success to the thickness of the slices, the length of time between the two immersions, or the temperature of the oil in each case.

Which is the relevant parameter? Why do the potatoes puff up? How can this puffing up be optimized? In testing the classic recipes one needs to keep two things in mind: that potato cells contain granules of starch, which swell when the cellular water is heated, forming a purée, and that because a potato is a thermally isolating material, its center is slow to cook. If the oil in the first round of frying is too hot, an excessively thick and rigid crust forms before the center is cooked, and the potato will not puff up.

Water Vapor Repels Oil

Next, if we weigh the fried slices, we find that the oil does not replace the water eliminated by heating, as was long assumed. Given a surface of 100 square centimeters (or roughly 15 square inches), about 80 cubic centimeters (almost 5 cubic inches) of steam manages to escape per second. In other words, the pressure of the steam keeps the oil from seeping in. Besides, if the slices quickly rise to the surface, this is because the water has been replaced by steam and not by oil (a potato is composed of 78% water and 17% starch, which is denser than both water and oil).

The behavior of steam bubbles provides the key to the phenomenon of *soufflage*. In order for the slices to puff up, steam must suddenly be generated, deforming the crust, whose dried-out cells create a steam-resistant compartment within each slice. When vaporization is slow, small trains of bubbles trickle out through openings in the crust, and the pressure of the steam is insufficient to cause the slices to expand, hence the need for hotter oil during the second round of frying.

Puffing up also requires that the compartments formed during the first round of frying be impermeable. The centers of the slices continue to cook during the interval between the first and second rounds, and water is redistributed through the dried-out areas. As the temperature falls the crust probably becomes detached from the center as well. The second round of frying then causes the residual water in each slice to evaporate, triggering expansion because the steam has a hard time escaping through openings in the compartment walls.

This explains why the thickness must be carefully controlled: If the slices are too thin, one does not obtain a crust with an intermediate layer of puréed starch granules, and the quantity of steam generated therefore is insufficient; if the slices are too thick, more time is needed for the center to cook and an overly thick crust forms on the outside, hindering the expansion. It also becomes clear why the greatest care must be taken in handling the potato slices. For if the thin crust is pierced, large vapor bubbles are suddenly able to escape through the openings, and the pressure is no longer sufficient to cause the slices to puff up.

Finally, how can the amount of oil absorbed by the puffed potatoes be minimized? Sam Saguy at the University of Jerusalem has shown that the oil is present mainly on the surface of the sliced potato, in quantities that increase with the rugosity of the surface and repeated use of the same oil: The more uneven the surface, the more oil that adheres to it (because of an increase in tensioactive molecules that results from repeated heating, hence the foam produced by old oil). It is a good idea, then, to fry potatoes in clean oil, to use as sharp a knife as possible, and to wipe off any oil coating the surface of the cooked potato slices so that when the water inside cools and condenses it is not absorbed.

15

Preserves and Preserving Pans

Why are unplated copper pans recommended for cooking fruit preserves?

LET'S EXAMINE A FEW MORE DICTUMS. L.-E. Audot, author of *La cuisinière de la campagne et de la ville* (1847), says that in order to make fruit preserves "it is indispensable to use an unplated copper pan (earthenware or terracotta ones being liable to burn [the preserves] or impart a bad taste)." Sixty years later, geologist Henri Babinski, in his *Gastronomie pratique* (1907), advised, "For preserves made from red fruits, it is preferable to use an enameled pan, which does not transmit any sharp taste, as often happens with unplated copper pans." During the same period, professors at the École du Cordon Bleu recommended that cooks "avoid using any iron or tin-plated utensil."

What is one to make of these conflicting opinions? Should copper or enameled cast iron be used? If copper, tin-plated or unplated? Although copper preserving pans may retain a certain luster that encourages culinary nostalgia and adds to the aesthetic quality of the kitchens in which they are displayed, they are also a bother because they have to be thoroughly cleaned (which cannot be done with ammonia, by the way, because this would give the preserves a disagreeable taste) before being used. Why not use stainless steel pans or enameled containers instead? Does copper give better results because of its superior thermal conductivity? Or does it possess other unsuspected properties that make it preferable to these alternatives?

The Role of Copper in Preserves

Nothing beats an experiment. Let's begin by putting red currants or raspberries in an unplated copper pan. To be rigorous about it, let's first measure the pH of the pan's contents (pH is a measure of acidity running from 0, for very strong acids, to 14, for very strong bases). The acidity of such fruits sometimes is surprisingly high. Indeed, a pH of about 3—which is to say about as much as certain vinegars—is not unheard of. Next, tilt the pan and you will see that the copper has been stripped away by the fruit and its juice. In other words, the copper ions covering the metal have dissolved.

Do these ions have an effect on the preserves? Let's conduct another experiment, dividing a previously cooked batch of preserves in a chemically inert container (glass, for example) and then adding a copper salt to one of the two halves. When the two portions have cooled, one observes that the one containing copper ions is firmer than the other. Why? Because the solidity of the preserves depends on the presence of pectin molecules, extracted from the fruits, which form a network that traps the water, sugar, and fruits. Adding lemon juice generally promotes gellification because pectin molecules contain carboxylic acid $-COOH$ groups that, depending on the degree of acidity, may or may not combine. If the environment is insufficiently acidic, the carboxylic acid groups are ionized in $-COO^-$ form so that the electrical charges they carry have a mutually repulsive effect; in an acidic environment, by contrast, these groups are neutralized and the pectin molecules no longer repel one another.

What is the role of copper in all of this? In preserves copper is found in the form of ions and possesses two positive electric charges that interact with the two negatively charged groups, causing the pectins to bond with one another. In other words, copper reinforces the pectin gels, hardening preserves, as experience shows.

And Tin?

Given that copper is a suitable material, why should the tin that covers the inner surface of old preserving pans be harmful? Could it be that the old dictums are nothing more than the worthless residue of empirical advances in culinary practice? As it turns out, putting red fruits such as raspberries or currants in tin-plated containers produces no unwelcome consequences. Because tin does not act on red fruits, one might suppose that copper is the culprit, but fruits placed in copper are not altered either.

It is nonetheless generally the case that metals act through their salts. Try sprinkling a pinch of various metallic salts—silver, aluminum, copper, tin, iron, and so on—over red fruits. The tin salts immediately cause a disagreeable purplish color to appear because the pigments in the red fruits have combined with the metallic atoms; and because the electrons responsible for this bond are differentially distributed in the molecules of the pigments, these pigments absorb light to different degrees. Thus silver salts cause raspberries to whiten a bit, whereas copper ions give them a fine red-orange color. Tin ions trigger the purple tinge that has given rise to the prejudice against tin-plated pans.

Modern cleaning methods are superior to those in times past, however, and so the dictum must be amended: Red fruits should never be placed or cooked in unclean tin-plated copper pans.

16

Saving a Crème Anglaise

A pinch of flour prevents the formation of protein aggregates, which nonetheless can be broken up in a mixer.

HOW CAN ONE SAVE A CRÈME ANGLAISE that has curdled? The question is of culinary importance because crème anglaise figures in one form or another in many desserts. One of the differences between crème anglaise and crème patissière, for example, is that although both are composed of milk, egg yolks, sugar, and an aromatic ingredient such as vanilla, crème patissière additionally contains a certain amount of flour, which protects against curdling. Crème anglaise, lacking this protective agent, is more liable to turn.

The Second International Workshop of Molecular Gastronomy, held in April 1995 at the Ettore Majorana Center in Sicily, brought chefs and physical chemists together to examine such questions. Two renowned chefs, Christian Conticini and Raymond Blanc, raised a series of issues relating to sauces and dishes derived from them: crème anglaise, crème patissière, mayonnaise, stiffened egg whites, soufflés, chocolate cream fillings, jellies, preserves, and so on. The physicists and chemists, led by Nicholas Kurti, Pierre-Gilles de Gennes of the École Supérieure de Physique et de Chimie Industrielles in Paris, and me, sought to identify the mechanisms for culinary effects that were plain to see but not yet understood scientifically, regarding sauces as solutions, emulsions, foams, gels, suspensions, and so on.

At the conference the problem of crème anglaise was examined experimentally. Like many other delicate dishes, crème anglaise has inspired many dictums and ad hoc remedies that are worth scrutinizing. It has long been said that a pinch of flour added to crème anglaise will keep it from curdling. Why should this be so? Is it true that a crème anglaise that has turned can be saved by vigorously shaking it in a blender?

The Pinch of Flour and the Blender

Crème anglaise was examined at different stages of preparation. First, the custard was gently heated (at a lower temperature of 65°C [149°F]); gradually, as in the case of every successful crème anglaise, the mixture of milk, sugar, and egg yolks thickens. Under the microscope, small-scale structures (a few micrometers long) can be seen.

Then this same custard was put in a microwave oven for a few seconds so that a curd—a sign of overcooking—would appear. The observed microscopic structures were about twice as large and dense as those that had been observed in the successful custard, but their general aspect was not substantially different. When the custard was overheated, its appearance under the microscope changed completely: Clear liquid areas separated very dense areas composed of structures similar to those that had been observed at the onset of curdling.

Finally, this botched crème anglaise was mixed for a few dozen seconds. To the naked eye it had become frothy. The curdles had disappeared, and the texture of a perfect custard seemed to have been restored; under the microscope, however, a state of aggregation intermediate between that of a perfect custard and that of the initial stage of curdling could be discerned.

Inevitable Coagulation

It seems clear that the setting of a crème anglaise depends on the coagulation of the egg yolk, which occurs whether the result is successful or not. The structures observed under the microscope probably are aggregates of proteins that have been partially broken down by the heat, then reassembled by means of weak chemical bonds.

When the crème anglaise has been excessively heated, coagulation is rapid and produces macroscopic aggregates, evidence of curdling, which can be eliminated by mixing. Has the possibility of curdling thereby been completely removed? Can a perfect crème anglaise be reconstituted by putting a botched one in a blender?

Microscopic analysis shows that although a blender does a good job of dissociating the macroscopic aggregates, more agitation is needed if the result is to contain only microscopic protein aggregates, similar to the ones found in a successful sauce. Those whose palate is sufficiently refined to detect the omelet taste characteristic of a botched crème anglaise can prevent the sauce from curdling in the first place by adding a pinch of flour before cooking. Curdling will not take place, even if the crème anglaise is boiled.

The reasons for this protection are still a subject of debate, but it is known that placing starch granules in a hot liquid triggers the release of amylose molecules and that the water penetrates the granules and causes them to swell. These swollen granules, together with the long, dissolved amylose molecules, limit the movement of proteins, blocking the formation of macroscopic protein aggregates.

17

Grains of Salt

Culinary myths and legends of the white gold.

MYTHS ABOUT SALT DIE HARD. For example, some cooks recommend adding salt only to water that is already boiling because salty water, they say, takes longer to boil than pure water. This rule is widely believed, but is it justified?

Or consider the question of boiling meat to make stocks. Many cookbooks say that meat should be salted first in order to better extract its juices. Is the promised efficiency real or illusory? On the other hand, salad is not to be salted too far in advance because seasoning it in this way will wilt the leaves. What is one to make of this dictum, which comes from Japan?

With regard to vegetables, the French chemist Michel-Eugène Chevreul recorded the following observations in *Recherches des matières fixes tenues en dissolution dans l'eau pure ou salée, qui a servi à la cuisson des légumes* (1835): "Water used for the decoction of vegetables had a reddish-brown color; it retained a sensible quantity of the odiferous principles. Salted water used for the cooking of the same vegetables had a more pronounced fragrance than pure water; its flavor, allowance being made for the flavor peculiar to salt, was also more pronounced, and yet, remarkably, it contained a lesser proportion of extractive matter." Is this account to be credited?

We should be wary of assertions that are not supported by sound experiments. With respect to the first claim, it is true that adding salt to water cools it down (because the water loses energy in dissolving the salt) and raises its boiling point and that the mass of fluid to be heated is greater. But whether salt is added before or after the water has been brought to a boil, all three phenomena are equally in play. If we heat water with and without salt (the same quantity of water, in the same pan, heated in the same way), we find that, within the limits of experimental error, the time needed to bring it to a boil is the same.

With regard to stock, it is commonly maintained that the major effect of salting the meat in advance is to promote osmosis. Heating causes the various molecules to move apart so that their final concentrations are everywhere the same, but because the larger molecules do not pass through the cellular membranes, it must be water that enters or leaves the cells, depending on the case. If one takes into account only this effect, one would predict as a theoretical matter that the meat must lose more juice when it is cooked in the presence of salt. Let us then conduct a simple but careful test. Put two identical pieces of meat in two identical pans, one containing pure water and the other water that has been salted to the saturation point. Now cook them for the customary five hours, weighing the two pieces every ten minutes. After the prescribed time has elapsed, the mass of the two pieces is the same, give or take a gram.

Would eggs react differently? It is sometimes said that the water used for cooking hard-boiled eggs must be salted because otherwise the water would infiltrate the eggs by osmosis, causing them to expand and to crack. Let's cook a dozen eggs in a pan full of pure water and a dozen eggs in a pan identical to the first, filled with the same quantity of water, only this time highly salted. Again, let's weigh the eggs at various stages of the cooking process to determine whether the suspected osmosis actually causes their weight to increase. Then we will count the number of cracked eggs in each batch.

Going to all this trouble is worthwhile because it reminds us of something that chefs knew in the nineteenth century but have forgotten since. Although salting the water does not prevent the shells from cracking (the best way to ensure that they don't is to pierce the eggs with a needle, which by permitting the air trapped inside the egg to escape more easily preserves the integrity of the shell), it seasons the white of the egg, imparting flavor to an otherwise tasteless material.

Wilted Salad

Let us now turn to the effect of salt on vegetables. How do salad greens react when they are sprinkled with table salt? It turns out that there is no reaction, even after several hours, as long as the lettuce is dry. This is because its leaves are covered with a waxy cuticle that prevents osmosis.

What about vegetables cooked in water? To determine how much matter is lost during cooking, let's experiment with onions and carrots, covering the pan and weighing them at intervals. Onions, it turns out, lose more of their mass when they are cooked for a short time in salted water. In this case the effects of osmosis are plain. With carrots, however, there is no great difference. In both cases the vegetables decompose more when they are cooked in salted water, but if the cooking goes on too long, they both wind up in the same degraded state whether or not salt has been added to the water.

Why do explanations that assume osmosis do a poor job of predicting the actual behavior of vegetables and meats cooked in water? A glance through the microscope gives the answer: Animal and vegetable cells are not covered with semipermeable membranes through which osmotic transfer can take place. Muscle cells are sheathed by collagen, and vegetable cells are protected by a rigid wall. Some degree of osmosis may occur at the beginning of cooking, but over time the structure of the cells is degraded, with the result that they open up and absorb fluid like a sponge. Salt doesn't change anything.

There are many other such dictums. Putting cooked and peeled potato in a sauce that is overly salty, it is said, will remove the saltiness. Some chefs claim to be able to see tiny shimmering particles in sauces that have been oversalted. In the fourteenth century, Guillaume Tirel (known as Taillevent) noted that soups and stews tend to boil over if salt and fat are not added to them. In the case of grilled meats it is often said that salt should not be added at the beginning of cooking but reserved until the end so that it will be soaked up by the meat. To remove extra salt from a dish, Ginette Mathiot, author of the bestselling *La cuisine pour tous* (1955), advised immersing a cube of sugar in it for two seconds. What do you make of these recommendations?

18

Of Champagne and Teaspoons

A teaspoon in the neck of a bottle of champagne does not prevent the bubbles from escaping.

A TEASPOON PLACED IN THE NECK of a bottle of opened champagne, it is said, will help preserve its fizz for a certain time. Some even go so far as to claim that the effect occurs only with silver spoons. What reason is there to believe such dictums, which seem to be the product of nothing more than unscientific experimentation? By virtue of what strange physicochemical principle could a teaspoon trap the bubbles of this noble beverage of celebration? As it happens, there is no need for us to inquire into this matter ourselves because the Comité Interprofessionnel du Vin de Champagne has already carried out a rigorous series of experiments on the "teaspoon effect." Their findings, which turned out to be negative, were initially published in the *Vigneron champenois*.

Michel Valade, Isabelle Tribaut-Sohier, and Frédéric Panoïotis had the advantage of being able to institute reliable controls that few of us have, particularly access to an ample supply of bottles of champagne from the same vintage, differing from one another as little as possible. To simulate consumption, the bottles were partly emptied, either by a third or by two-thirds. Some were left open; others were equipped with the famous teaspoon, silver or stainless steel depending on the case; still others were stopped with a cork; and a final batch was sealed by a metal crown cap. All the bottles were placed in an upright position at a temperature of 12°C (54°F). In order to test the quantity of residual gas, pressure, loss of weight, and taste were measured at regular intervals.

Pressure Variations

The pressure of the champagne at the outset was about 6 bars. As the bottles were emptied, the pressure fell to 4 bars when the residual volume was 50 centiliters (16.9 oz.) and to 2 bars when the volume was 25 centiliters (8.5 oz.). Subsequently, the drop in pressure was similar for all the bottles that had been left open, with or without a teaspoon; it was less for the bottles with caps (10%, instead of 50% in the case of bottles left open for forty-eight hours).

To forestall criticism of these preliminary measurements from skeptics attached to the myth of the teaspoon, the researchers also measured the loss of weight through degassing. Once again the weight loss was observed to be identical for the open bottles and for those whose neck was fitted with a teaspoon. By contrast, the loss was zero for the bottles that had been stopped with a cork. Although a little gas escapes from the liquid in this case, its accumulation on top of the liquid limits any further release; when the bottle is uncorked the next time for measurement, the noise of its being opened reveals the presence of this accumulated gas, but the quantity of dissolved gas remains greater because the cork stopper limits the degassing of the wine.

In Vino Veritas

These measurements confirmed that degassing depends chiefly on the pressure exerted downward on the liquid, the presence of suspended matter in it, and imperfections in the inner surface of the bottle. This last effect is clearly visible when one puts sand in a glass of champagne or when champagne is poured into a frosted glass: bubbles immediately form in great numbers, triggered by the irregularities introduced in the liquid.

Saving the best for last and for themselves, the Epernay researchers tasted the contents of the various opened bottles. This blind tasting confirmed that the teaspoon did nothing to preserve the sparkling character of the champagne. By contrast, the samples that had been preserved in hermetically sealed bottles were more effervescent. In every case the wines had oxidized because oxygen had been let in when the bottles were partly emptied. But never mind that cork stoppers preserve champagne better than teaspoons; one should not put off until tomorrow what one can do today. Once you've opened a bottle, finish it off!

19

Coffee, Tea, and Milk

Determining the most efficient way to cool down a drink that is too hot.

THE MOST COMMON EVERYDAY EXPERIENCES give rise to practical questions. Is it true that running, rather than walking, under the rain will keep you drier? Does it really feel cooler to wear white clothes on a sunny day? Does water always drain out from a bathtub in a clockwise direction in the Northern Hemisphere? Testing such questions experimentally is a simple matter, but we are rarely willing to go to the trouble.

Small mysteries of this sort abound in cooking. Take the coffee we drink every morning. It is always boiling hot, and we are never sure how to cool it down. Some great minds have taken an interest in the question, including the British physicist Stephen Hawking. According to Hawking, if you want to avoid burning your mouth you should wait a few moments before adding sugar, assuming you sweeten your coffee, because it will cool down more rapidly than if you add the sugar right away.

What is the theoretical basis for this prediction? Stefan's law, which stipulates that the heat radiated by a body per unit of time is proportional to the fourth power of its absolute temperature, tells us that hot bodies radiate more energy than cold bodies. In allowing the coffee to cool down by itself one profits from its higher temperature, and therefore from its more intense radiation; dissolving the sugar afterward completes the process. Putting the sugar in first, on the other hand, has the effect of immediately cooling the coffee, so less advantage is derived from its radiation. This is plausible enough, but is it true?

Theory to the Test

Experiments (all the ones mentioned here were done under controlled conditions that very closely reproduce actual experience) show that the effect predicted by Hawking is too weak to be observed. Moreover, they reveal that the cup plays an important role: As cooling begins, the cup is reheated, but because it gives off heat only by conduction its temperature falls less rapidly than that of the liquid, whose cooling it subsequently retards.

What if one were to add cold milk instead of sugar? Once again the experiment could hardly be simpler: Make a very hot cup of coffee, then add milk at room temperature and measure the rate of cooling. Next, compare this result with what happens when a very hot cup of coffee is allowed to cool beforehand and room temperature milk added at a later stage.

This time the theoretical prediction is correct. For 20 centiliters (6.75 oz.) of boiling coffee that is cooled initially with 7.5 centiliters (2.5 oz.) of room temperature milk, a comfortable drinking temperature of 55°C (131°F) is obtained after about 10 minutes, but if one waits for the coffee to reach 75°C (167°F) before adding the milk, one obtains the same result after only four minutes. Knowing an elementary law of physics reduces the wait by more than half.

Teaspoon and Radiator

Where does this leave those who take neither sugar nor milk? Should they put a teaspoon in their coffee, reasoning that the heat of the coffee will travel through the metal of the spoon and thence radiate into the atmosphere?

This time let's skip theoretical calculations, however interesting they may be (for the first time in my life I was able to use Clairaut's equation, which as a student I had worked so hard to learn) and use a thermocouple to measure the actual effect instead. We will see that the effect of the teaspoon is virtually nil: When one takes two cups containing coffee at an initial temperature of 100°C (212°F) and puts a teaspoon in one of them, the difference in temperature after ten minutes is less than 1°C (or 1.8°F). A teaspoon is not an efficient radiator, even when it is made of silver.

Coffee cools slowly, but what about tea? The difference in color between coffee and tea affects the degree of radiation in principle, but in practice it does not affect the rate of cooling because the effect is too small to be experimentally observed: Assuming identical cups and identical quantities of liquid, the cooling curves coincide exactly. On the other hand, a large bowl of tea may cool more slowly than a small cup of tea, but that is another story.

Let's conclude by examining the intuitive behavior of drinkers of hot beverages everywhere. Is blowing on coffee an efficient way of cooling it? What about stirring it with a spoon? Experiment shows that a boiling liquid that cools spontaneously by 6°C (11°F) per minute cools by 11°C (20°F) per minute when one stirs and one blows on it at the same time. Which is more effective: stirring or blowing?

Stirring has two effects: It makes the temperature of the liquid throughout the cup uniform and, by bringing the hottest part up to the surface, accelerates cooling. Stirring therefore increases the area over which energy is exchanged between the hot liquid and the cold air. A further consequence of this ventilation is that it expels from the liquid the fastest-moving molecules, which have now entered into the steam phase. This means that they are not recycled through the rest of the liquid, with the result that its average molecular kinetic energy is reduced. Because temperature measures exactly this, the average velocity of molecular agitation, a lower velocity means that the liquid has cooled. Let's blow vigorously on the coffee in one cup while vigorously stirring the contents of another cup, trying in the latter case to maximize the surface area of contact between air and liquid, as the theoretical description of the problem requires.

Question: What will we observe? Answer: Blowing is much more efficient than stirring. The same coffee that loses 6°C (11°F) per minute when one blows on it cools down only by 3.5°C (6°F) when it is stirred.

2 The Physiology of Flavor

PART TWO

20

Food as Medicine

Our primate cousins vary their diet depending on their state of health.

TO THE GREEK MIND, barbarians were people who did not change their diet when they were sick. At the Laboratoire d'Écologie Générale of the Museum in Brunoy, Claude Marcel Hladik and his colleagues demonstrated that under certain circumstances monkeys eat earth, or plants containing alkaloids, in order to preserve a balanced diet—even to treat intestinal disorders. In this they are close relatives of the civilized world.

The reaction of primates to sweet solutions is a striking feature of mammalian evolutionary adaptation: The larger the animal, the more efficiently it detects sugars. Large animals, better equipped to recognize sweet foods, are able to acquire more energy for themselves. An exception to this rule, which holds as much for fruit-eating animals as for herbivores, is the loris *Nycticebus coucang*, a prosimian that is insensitive to sucrose (ordinary sugar). This may be because it must tolerate the bitter taste of insects in addition to various prey that other monkeys do not feed upon.

A Normal Primate

Regarded in terms of their ability to perceive sweet products, humans are normal primates: Our body mass is large, and we are very sensitive to sweet tastes. Nonetheless, this biological basis is modulated by environmental factors. For example, perceptual thresholds for glucose and sucrose differ between the inhabitants of tropical forests and those of grasslands. Thus the Pygmies, who occupy forests where sugar-rich fruits are common, have a less developed sensitivity to sweet tastes than that of peoples who live in savannas, where plants contain less sugar.

Tests of species sensitivity to sweetening agents have brought out surprising differences. For our species, monellin (a protein present in the red berries of the African shrub *Dioscoreophyllum cumminsii*) is 100,000 times sweeter than sucrose. Nonetheless, although the taste of this protein is identified by African nonhuman primates, it is not perceived by American primates. The same difference is observed in the case of thaumatin, a protein sweetener extracted from the fruit of another African plant, *Thaumatoccus danielli*. It may also be encountered in the case of brazzein, identified in 1994 in the creeper *Pentadiplandra brazzeana*.

Differentiation of receptor proteins in the papillary cells of the tongue probably occurred 30 million years ago, after the separation of the New World Platyrrhina and the Old World Catarrhina. In their respective environments these animals found various plants with which they evolved in tandem, eating the fruits of these plants while dispersing their seeds. In the Americas, where no protein sweetener has yet been found, coevolution should have caused new molecules to appear that would not have seemed sweet to Old World monkeys.

It has been known for several decades that vertebrates are able to detect sodium chloride and actively seek it in case of insufficiency. For example, horses lick salt deposits only if they have to, an observation confirmed by the study of salt-deprived rats. In natural environments (particularly forests) salt deficiency is rare, but in 1978 the American biologist John Oates observed that *Colobus guereza*, a shy monkey that seldom ventures out of its normal tree habitat, comes down to the ground to eat the leaves of the plant *Hydrocotyle ranunculoides*, which grows in ponds and contains more salt than other available sources.

Natural Medicine

Other primates eat earthy matter even though they do not suffer from salt deficiency. The soil eaten by *Colobus satanas* (another colobus monkey that lives in the forests of Gabon) contains less salt than the fruits that make up an important part of its normal diet, but this behavior occurs during the two periods of the year when the animal must supplement its diet with mature leaves (feeding the rest of the year on young shoots and leaves in addition to flowers, fruits, and grains). These older leaves contain not only molecules of the polyphenol family (hydroxyl [–OH] groups that attach to benzene rings having six carbon atoms) but also tannins, which inhibit the digestion of proteins by forming complexes with them. Because clay and other soil compounds readily absorb tannins, the geophagy of these monkeys can be explained as a way of compensating for the ingestion of unwanted plant products. "From your food you shall make your medicine," Hippocrates is credited with saying. Could it be that our primate ancestors whispered this phrase in his ear?

Often an aversion to bitter tastes favors the avoidance not only of dangerous alkaloids but also of astringent compounds such as tannins, terpenes, saponins, and strong acids. Nonetheless, not all toxic compounds are bitter: Dioscin, a lethal alkaloid found in the yam *Dioscorea dumetorum,* is almost tasteless. The animal kingdom is protected by the phenomenon of neophobia—the fear of eating what is new—and by conditioning from postingestive symptoms that trigger the appearance of an aversion (observed in rats and primates alike). Even so, chimpanzees are known to heal themselves by eating the bitter plant *Vernonia amygdalina,* generally avoided by healthy animals. This plant contains several steroidal glycosides that are effective in treating gastrointestinal troubles. What shall we call this type of behavior? Natural medicine?

21

Taste and Digestion

The absorption of monosodium glutamate triggers mechanisms for assimilating proteins.

WHY DO WE STOP EATING even though only a small quantity of metabolites has entered the bloodstream? The sensation of a full stomach does not signal satiety: A rat whose belly has been pumped full of air does not cease to eat. Through a series of reflexes, however, the organism is able to anticipate the metabolism of foods. For example, a bit of sugar placed on the tongue triggers the almost immediate release of glucose by the liver.

Around 1960, Stylianos Nicolaïdis and his colleagues at the Collège de France observed that the stimulation of the taste receptors by saccharine caused two hormones to be released by the pancreas: glucagon, which is responsible for the release of glucose, and insulin, which is responsible for the metabolism of glucose. Moreover, stimulation of receptors for the sweet taste produces an anticipated reaction that enables the body to metabolize glucides.

More recently, Nicolaïdis and a team of researchers including Claire Viarouge, Patrick Even, and Roland Caulliez examined reflexes that are triggered by detection of the taste of proteins and prepare the organism to metabolize them. One aspect of their investigation involved the ingestion of monosodium glutamate, which earlier studies had suggested is perceived by the organism as a signal to begin absorbing proteins. Previously limited to the seasoning of oriental soups, monosodium glutamate is now very commonly used in the food industry as a flavor enhancer. In addition to a salty taste, for some people it possesses a particular taste called umami that is distinct from the four classically recognized tastes (salt, sweet, sour, and bitter); mind you, the classic theory is wrong, as all good neurophysiologists know. Two questions presented themselves. First, does monosodium glutamate increase metabolic intensity in the same way that the absorption of proteins does? Second, does it trigger the same reflexive release of hormones (glucagon and insulin) as the one normally induced by the appearance of amino acids in the bloodstream?

To study these questions the team at the Collège de France constructed a device capable of isolating the various components of the metabolism and separately recording the metabolic demands associated with locomotion and thermogenesis, or heat production, in relation to food intake. Thermogenesis measures metabolic intensity without regard for an animal's locomotive behavior. Experiments performed in 1991 to determine levels of thermogenesis during periods of activity and in the resting animal showed that stimulation of rats by intravenous administration of monosodium glutamate in the mouth and stomach produced only a weak hormonal reaction. Although it induced an anticipatory reflex, the monosodium glutamate seemed to be effective only when it was associated with other signals characteristic of food intake.

To test this hypothesis, the physiologists placed cannulas in the mouths of rats in order to be able to directly inject solutions of monosodium glutamate or distilled water using a catheter connected to the top of their cage. The rats were free to move around, and their metabolism was continuously monitored.

Immediately after implantation of the cannulas, the rats were trained to receive water or monosodium glutamate solutions through these tubes. One group of rats was injected with various concentrations of monosodium glutamate while they were eating. The other group, used as a control, received only distilled water during this time.

The physiologists were able to verify that the taste of sodium, when it accompanies a complete meal, causes notable metabolic changes. The thermogenesis induced by food intake increased much more, and much more rapidly, in the rats injected with monosodium glutamate than in the control group, suggesting that monosodium glutamate acts as a sort of protein "saccharine" capable of misleading the organism by its taste. Although it consisted mostly of carbohydrates, the meal was perceived as primarily protein.

How long the organism allows itself to be fooled when this gustatory information is not confirmed during the course of metabolism and whether the anticipatory reflexes contribute to satiation are questions that remain to be answered.

22

Taste in the Brain

The cerebral areas activated during the perception of tastes have been identified by means of nuclear magnetic resonance imaging.

JEAN-ANTHELME BRILLAT-SAVARIN OBSERVED in Meditation 2 of *The Physiology of Taste,* "Up to the present time there is not a single circumstance in which a given taste has been analyzed with stern exactitude, so that we have been forced to depend on a small number of generalizations, such as *sweet, sugary, sour, bitter,* and other like ones which express, in the end, no more than the words *agreeable* or *disagreeable,* and are enough to make themselves understood and to indicate, more or less, the taste properties of the sapid body which they describe. Men who will come after us will know much more than we of this subject; and it cannot be disputed that it is chemistry which will reveal the causes or the basic elements of taste."

Brillat-Savarin was prescient. Today biochemists and neurobiologists know a good deal about the function of receptor molecules for tastes, which are located on the surface of papillary cells on the tongue. Even so, nuclear magnetic resonance imaging (MRI) techniques are helping them better understand how information perceived by the taste receptors travels up into the brain for processing.

With the development of these techniques, which record cerebral activity by detecting changes in blood flow in the brain, neurobiologists have paid special attention to cognitive activities such as language, calculation, and memorization. Olfaction has been less studied and the perception of tastes completely neglected. Barbara Cerf and Annick Faurion of the Laboratoire de Neurobiologie Sensorielle in Massy and Denis Le Bihan of the Centre Hospitalier d'Orsay in Paris have identified the cerebral areas activated by taste molecules.

Basic knowledge was rudimentary. The only thing that was known, from observation of people whose brains had been partly destroyed by wartime injuries, was that the parietal operculum, located near the central sulcus (Rolando's fissure), undoubtedly played a role in the perception of tastes. Nonetheless, electrophysiological studies yielded contradictory results, which pointed instead to another area located in the insula.

Because an MRI requires subjects to lie down inside a tunnel-like machine, they were fed with solutions transmitted through flexible tubes. These constraints determined the sapid substances that were tested: The subjects received solutions of aspartame (a sweetener), sodium chloride, quinine (bitter taste), glycyrrhizic acid (licorice taste), guanosine monophosphate (the umami taste, similar to that of monosodium glutamate, used in Asian cooking), and D-threonine (indescribable—you have to taste it for yourself). The experimenters first gave the subjects water, then sapid solutions, then water again, and so on, in order to forestall habituation while sustaining stimulation for several dozen seconds, the time needed for the MRI device to record a signal.

Lateralization of Taste

The subjects who received these solutions were instructed to concentrate on their taste in order to minimize interfering activations of other parts of the brain. The subjects described the intensity of their sensations by moving a cursor along a graduated scale. By calculating correlations between the various perceptual profiles and activations of the different areas of the brain, the neurobiologists were able to determine which activations were linked to the perception of tastes. Individual differences were pronounced and the images noisy, so that many experiments had to be analyzed in order to pinpoint the areas that were specifically associated with the perception of taste.

The first studies showed that four cerebral areas are activated by sapid solutions: the insula and the frontal, parietal, and temporal opercula. There is no single taste center in the brain nor any cerebral areas that are specifically linked to particular tastes. On the other hand, certain areas that were not systematically activated are known to play a role in language comprehension, hence the hypothesis that the detection of taste may be associated with the act of naming it.

A second study comparing five left-handed and five right-handed people found that the four cerebral taste areas were not systematically inverted between the two groups. By contrast, activation of the insula differed according to handedness. This area is composed of two regions. The upper part is activated in both hemispheres, for right-handers as well as left-handers; the lower part is activated unilaterally in the subject's dominant hemisphere. The perception of taste therefore is lateralized in a way that is analogous to language use and motor activity.

A third series of studies compared subjects' reaction to molecules that have only taste and to molecules that have both taste and astringency (or pungency). This time the activated areas were analogous, which explains why flavor—the overall sensation that is registered in the course of eating—is so all-encompassing and so difficult to describe: The brain constructs a global sensation through the synthesis of signals coming from various types of receptors.

23

Papillary Cells

The functioning of the cells that allow us to perceive the taste of foods is discovered.

IN 1994, RICHARD AXEL AND LINDA BUCK at the College of Physicians and Surgeons of Columbia University in New York announced the discovery of proteins in the membrane of nasal cells that capture odorant molecules and make olfaction possible. The news caused a great stir, but it failed to satisfy researchers interested in the related problem of taste. Two years later another team of biochemists at Columbia, Gwendolyn Wong, Kimberley Gannon, and Robert Margolskee, published the results of a study of gustducin, a protein found in the papillary cells of the tongue that had been cloned in 1992 but whose function was unknown. They observed that inhibiting the synthesis of gustducin in the taste cells of mice caused these animals to lose their aversion to bitter flavors and, still more surprisingly, their sensitivity to sweet molecules.

Taste begins when a sapid molecule binds with receptors or ion channels in the membrane of a papillary sensory cell. Once the electrical potential of the papillary cell is sufficiently modified as the result of a series of reactions, the cell commences to excite neurons, which, little by little, convey information to the brain.

Not all taste molecules act in the same fashion. Whereas hydrogen ions (sour taste) and sodium ions (salt taste) act directly on the channels of taste cell membranes, immediately modifying the electrical potential of the cell by adding their electrical charge to its total charge, compounds of sweet, bitter, and other tastes (licorice, for example) bind to molecules known as receptors—no doubt proteins—that are located in the cell membrane, in contact with the extracellular environment.

It is thought that these receptors are paired with other previously identified proteins, the G-proteins, which trigger the emission of molecules known as second messengers that act within the cell. Nonetheless, the receptors are evanescent because they form only a weak bond with taste molecules. This is inconvenient from the scientific point of view, but, gastronomically speaking, it is an advantage: If taste molecules formed too strong a bond with receptors, we would not perceive the rapid succession of tastes in a dish.

Before the discovery of these receptors, Margolskee and his colleagues had begun investigating G-proteins, which are particularly abundant in the taste cells of the tongue. Using a method of genetic amplification involving the polymerase enzyme, they multiplied the number of genes of the alpha-subunits of several G-proteins, notably gustducin, which is uniquely expressed in such papillary cells.

These studies confirmed similarities between taste and vision. Gustducin was found to resemble a class of G-proteins found in the receptor cells of the eye known as transducins. Moreover, the Columbia team detected the transducin that is specific to the cones and rods of the eye in taste receptor cells.

The Eye and the Papilla

The resemblance proved to be enlightening. For if papillary cells function like the cells in the eye, then gustducin and transducin activate an enzyme that diminishes cyclic adenosine monophosphate production. In this hypothesis, the shortage of this second messenger would either modify the ion channels of the cell membrane and associated enzymes or disrupt the exchange of calcium ions between the inside and outside of the cell.

To test this hypothesis, the Columbia team inactivated the gene that codes for the alpha-subunit of gustducin and studied the behavior of mice born with this inhibited gene when they were offered various sapid solutions to drink. At the same time the neurobiologists recorded the electrical signals from the chorda tympani, a branch of the facial nerve that conveys gustatory information to the brain. The reactions were normal for salt and sour flavors but much weaker for bitter compounds such as quinine sulfate and denatonium benzoate and for sucrose (ordinary cane sugar) and a normally very intense synthetic sweetener.

Why was the perception of bitter and sweet not completely nullified? The neurophysiologists reasoned that because transducin plays a role, along with gustducin, in the perception of these tastes, the fact that it was not eliminated meant that they continued to be perceived, albeit to a lesser degree. Therefore their next experiment will investigate the consequences of inhibiting the genes for both transducin and gustducin.

24

How Salt Affects Taste

Salt transforms and softens bitter and sweet flavors.

TRUE GASTRONOMES HAVE TWO GREAT FEARS: gout and a diet without salt. To guard against gout they abstain, at least occasionally, from gamey meats; but against a salt-free regime they find themselves powerless and dread the doctor who prescribes it. This fear is doubly well founded. Gary Beauchamp and his colleagues at the Monell Chemical Senses Institute in Philadelphia have shown that the absence of the salt taste is not the sole inconvenience of this regime. Without salt, agreeable tastes forfeit their prominence, and they are unable to prevent disagreeable tastes from asserting themselves.

In earlier chapters I examined the action of salt on the texture of foods without discussing its taste. Salt is important because it increases the ionic strength of aqueous solutions, making it easier for odorant molecules to separate themselves from food. This is why unsalted soup has no flavor and why adding salt amplifies its odor, which is an important part of flavor. Sodium chloride is also a taste molecule that stimulates the papillary receptors. Does it have other virtues from the point of view of flavor? Does it really bring out the flavor of a dish, as some maintain?

In examining these questions experimentally, Beauchamp and his colleagues did not limit themselves to sodium chloride but also tested other salts such as lithium chloride, potassium chloride, and sodium aspartate. They sought to make sense of a paradoxical state of affairs: Whereas most psychophysiological studies test pairs of tastes and succeed in showing that salt either suppresses the accompanying taste or has no effect on it, every gourmet knows that unsalted foods lose much of their interest. Cooks who add salt to their pie dough—even a pinch in the case of a sweet pie—do so not in order to make the dough salty but to give it flavor.

Filtering Tests

The Monell Institute team of psychophysiologists wanted to know whether salt selectively filters tastes, weakening unpleasant tastes while enhancing pleasant ones. Convinced that it was not enough to examine pairs of tastes, they compared aqueous solutions containing one or more of three substances: urea (bitter), sucrose (table sugar), and sodium acetate. There were reasons for choosing these three: Sucrose added to urea softens its bitterness, and sodium acetate contributes sodium ions without imparting too salty a taste. Ten subjects were asked to evaluate the intensity of bitter, sweet, and other sapid sensations produced by combinations of urea, sugar, and salt in different concentrations (three for urea and salt, four for sugar).

As predicted, sodium acetate reduced the bitterness of urea. What gastronomic empiricism did not predict, however, was that salt masked the bitterness much more effectively than sugar. Mixtures of sugar, urea, and salt turned out to be sweeter and less bitter than unsalted mixtures of urea and sugar. Moreover, in strong sugar concentrations, the sweet character was increased by the addition of sodium acetate, probably because salt offsets the weakening of the sweet intensity caused by the bitterness of urea. Consistent with the hypothesis, the addition of sodium acetate by itself to sugar, in the absence of urea, did not increase the intensity of the sweet taste.

These studies were conducted for many other compounds and showed that sodium ions selectively suppress bitterness (and probably other disagreeable tastes as well) while intensifying agreeable tastes. It is therefore a question not of bringing out a single basic taste but rather of modifying the proportions of a combination of tastes. Adding salt to a variety of dishes—vegetables (both bitter ones, such as endive, and sweet ones, such as carrots and peas), certain fatty foods, and meats—may have become habitual because there is an unconscious wish to eliminate unpleasant tastes and to reinforce the natural sweetness of many foods. The recent experiments seem also to explain why some coffee lovers put a pinch of salt in the filter: to remove the bitterness of caffeine.

It is not yet known how the stimulation of taste receptors produces these effects, but we do finally know why salt-free diets make us wince.

25
Detecting Tastes

Discovery of a molecular receptor for a fifth taste.

ONE OF THE HOLY GRAILS OF PHYSIOLOGY is finally in our hands. For decades physiologists sought to explain how the cells of the gustatory papillae detect taste molecules. It was supposed that the surface of these cells contains proteins called receptors, to which the taste molecules attach themselves, but these receptors proved to be elusive. Attempts to extract them from papillary cells in solution were unsuccessful because receptors form a weak bond with taste molecules. Compensating for this experimental difficulty is a physiological advantage: If the bond were strong, receptors would be stimulated for long periods of time by a single molecule, and the resolution of individual tastes would be low. In that case gastronomes would be forced to savor their food in slow motion. Focusing on the phenomenon of weak molecular bonds, in February 2000 physiologists at the University of Miami identified one of the sought-after receptors, associated with the taste called umami.

It was long believed that the mouth is capable of detecting only four tastes, but the matter had been examined only cursorily. In 1908, Kikunae Ikeda at the Imperial University of Tokyo established that glutamate (the ionized form of an amino acid, glutamic acid) produced a particular sensation that was neither salt, sugar, sour, or bitter. After decades of struggle against conventional wisdom the notion of a fifth taste came to be accepted, in large part on account of the growing popularity in Western countries of Asian cuisine, which uses a great deal of monosodium glutamate. Moreover, it was shown that even animals detect this taste, perhaps because glutamate is present in many foods that are rich in proteins (which are chains of amino acids), such as meat, milk, and seafood. The detection of tastes is important because it signals satiety. One does not cease eating because one's stomach is full; one stops because the brain, alerted by the sensory system, notifies the organism that a sufficient quantity of food has been consumed.

From Mouth to Brain

In searching for glutamate receptors, Nirupa Chaudhari and his colleagues at the University of Miami took as their starting point the results obtained ten years earlier by Annick Faurion at the Laboratoire de Neurobiologie Sensorielle in Massy. Because glutamate is a neurotransmitter, which is to say a molecule that is exchanged between neurons in the brain (on being released by one nerve cell it binds to a receptor on the surface of a neighboring nerve cell), it seemed reasonable to scrutinize taste cells in the mouth for molecules analogous to neuronal receptors.

Building on Faurion's insight, Chaudhari and his colleagues looked for receptors paired with G-proteins—that is, with proteins embedded in the taste cell membrane that transmit the message detected by the receptor, which, projecting from the cell surface, is exposed to the extracellular medium. They made a surprising discovery: a truncated form of a neuronal protein known as a metabotropic glutamate receptor, or mGluR4.

Searching for Receptors

In the brain this neuronal receptor is activated by very weak glutamate concentrations. Unmodified, it would be completely saturated by the large quantities of glutamate present in the mouth when we taste certain dishes. Chaudhari and his colleagues concluded that the abbreviation of the protein's structure probably was an adaptation to the function of taste. Examining the receptor that is synthesized in the gustatory tissues of the rat, they found that it had lost the first 300 amino acids found in its neuronal counterpart and that in this truncated form it binds very weakly to glutamate. Moreover, the glutamate concentrations needed to activate it are similar to the perception thresholds measured in rats (a threshold perception being defined as the weakest concentration to which an animal is sensitive).

Why does the shortening of the protein's amino acid sequence reduce its affinity for glutamate? The University of Miami physiologists observed that the taste molecule strongly resembles a particular bacterial protein, extensively studied by crystallography, that has two bonding sites for glutamate. Truncation seems to have eliminated the more sensitive one of the two.

Is the protein they discovered the receptor for the umami taste? After all, the possibility could not automatically be excluded that the new protein transmits neural (rather than sensory) information from the gustatory cells to the brain. Confirmatory evidence is substantial. First of all, the mGluR4 protein is activated by other sapid molecules that rats do not distinguish from glutamate. This was not surprising because Faurion had observed that the tongue and brain react similarly to these molecules. Moreover, the truncated mGluR4 protein is synthesized only in the gustatory papillae. Finally, neither the complete nor truncated form of the mGluR4 protein has been detected elsewhere in the mouth than in the gustatory papillae.

This discovery raises many questions. Is truncated mGluR4 the only receptor for the umami taste? Is it involved in the perception of other tastes? Do receptor cells each bear only a single type of receptor? The result obtained by Chaudhari and his team seems to be the first of a series; physiologists have found other molecules of the same class that also seem to be taste receptors. Knowing the structure of the mGluR4 protein makes it possible to use molecular modeling systems to determine which molecules attach themselves to it. The computer-assisted design of completely new sapid molecules, further expanding the number of tastes, may not be far off.

26

Bitter Tastes

Not only are there more than four tastes, but several types of bitterness have been discovered.

IN 1995 THE PHYSIOLOGY OF SMELL took a great step forward with the identification of the proteins that constitute the receptors of nasal olfactory cells. However, the receptors of the papillary cells in the tongue and mouth that are sensitive to taste remained unknown. These gaps are gradually being filled. Alejandro Caicedo and Stephen Roper at the University of Miami have shown that the human gustatory system is capable of distinguishing several sorts of bitter taste.

In 2000, the same biologists who five years earlier had discovered families of olfactory genes found a vast family of receptors for what was then thought of as a single bitter taste. Other teams went on to show that individual receptors react selectively to compounds of a particular sort of bitterness. In parallel with this research it became apparent that each taste receptor cell expresses several RNA messengers, which in turn code for several receptors. It seemed natural, then, to suppose that individual receptor cells react to two or more taste compounds.

Neurological and behavioral studies conducted with rats, monkeys, and human subjects indicated that these different species are able to distinguish between several bitter stimuli. What is the cellular basis for this ability? Do taste receptor cells react specifically to one stimulus or to several? Progress in answering these questions was slow because ongoing neurophysiological investigation pointed to a variety of sources of interference along the nerve pathway that leads from the gustatory papillae to the brain.

Caicedo and Roper attacked the problem with a new imaging method that showed, in situ, the activation of taste receptor cells by a stream of calcium ions. A coloring agent sensitive to such ions was injected into the cells using a micropipette, and the reactions of the cells to various stimuli were observed while the distribution of these calcium ions was measured in real time with the aid of a laser, which excites the coloring agent, and a confocal microscope, which makes it possible to observe deep tissue cells. The University of Miami biologists were thus able to examine the reaction of several hundred taste receptor cells to a series of bitter compounds.

Cycloheximide was found to trigger strong but transient variations in the concentration of calcium ions in taste receptor cells. The four other molecules tested—denatonium benzoate, sucrose octaacetate, phenylthiocarbamide, and quinine—produced weaker but prolonged reactions lasting several minutes. The intensity of the reaction in each case depended on the concentration of the molecules to which the cell was exposed, which varied for each type of molecule. What is more, the recorded results corresponded to the behavioral response of rats fed with solutions of these bitter molecules (which act only above a certain threshold concentration).

Five Types of Bitterness

After meticulous study the Miami researchers established that only 18% of the 374 cells tested reacted to one or more of the five bitter compounds when they were administered at moderate concentrations. Among the cells that were sensitive to bitter compounds, reactions varied: 14% of the cells reacted to cycloheximide, 4.5% to quinine, 3.7% to denatonium benzoate, 2.4% to phenylthiocarbamide, and 1.6% to sucrose octaacetate.

The total proportion of taste receptor cells that register bitterness in the gustatory papillae (which contain other kinds of cells as well) turned out to be comparable to the proportion of cells sensitive to bitterness in the restricted population of taste receptor cells tested. None of the papillae studied seems specific to bitterness, and both the proportion and the distribution of cells sensitive to bitterness are comparable to those of the RNA messengers of receptors for bitter molecules. The Miami studies showed once again, only this time at the cellular level, that the different parts of the tongue are not specific to particular tastes, contrary to a view widely held among cooks and gourmets.

Applying the five stimuli to each cell in turn, Caicedo and Roper observed that a majority of the cells sensitive to bitterness reacted only to one of the five compounds tested, with their neighbors reacting to different ones. One quarter of these cells were activated by two bitter compounds, and 7% reacted to more than two of the five compounds. Note that these reactions were independent of one another: Stimuli were not simultaneously administered to a given taste receptor cell, and higher concentrations of bitter molecules did not increase the proportion of bitter-sensitive cells.

The sensitivity of individual cells to specific kinds of bitterness—obviously it will no longer do to speak of bitterness as though it were a single thing—would explain the observed behavioral reactions and, in particular, the capacity to make sensory distinctions between different sorts of bitterness. The nerve fibers that go out from cells specific to a given kind of bitterness appear to be grouped in dedicated bundles that communicate with a particular area of the brain.

It remains to come up with names for the various bitter tastes that are now known to exist.

27

Hot Up Front

Why spicy foods burn the mouth.

IN SEEKING TO COMPOSE a perfectly balanced and flavorful dish, the cook naturally looks to old recipes for ingredients whose combination has been tested and validated over the course of centuries. But traditional ways are not always the best. The various elements assembled from culinary experience must harmoniously stimulate not only the senses of taste and smell but also thermal and mechanical sensors, to say nothing of the chemical sensors for spiciness.

Our chances of success will improve if we have a precise understanding of the underlying molecular mechanisms. We know that sour and bitter tastes are offset by sweet ones and that salt facilitates the perception of other tastes, but the molecular basis of the physiology of taste remains incomplete. For example, why do hot peppers of the *Capsicum* family set the mouth on fire? Why are people who regularly eat hot peppers able to tolerate doses that novices find intolerable? Why do we like to eat foods that cause us pain? David Julius and his colleagues in the School of Medicine at the University of California, San Francisco (UCSF), cast light on these questions by studying the receptor for capsaicin, the active principle in chili peppers, paprika, and cayenne.

Pain and the Brain

An important advance in human physiology was made several decades ago with the exploration of the effects of morphine on the brain and the identification of the receptors for morphine and its derivatives. If an organism contains such receptors, researchers reasoned, molecules analogous to morphine ought to be able to be found in the body. This assumption turned out to be correct: Endogenous opioids were soon discovered, along with the regulatory system for suppressing pain.

Julius and his colleagues reasoned in a similar fashion that if our species consumes spicy foods whose molecules activate pain pathways, evolution ought to have endowed the human organism with receptors for endogenous molecules involved in signaling pain.

How to go about identifying these receptors? The UCSF biologists first isolated the RNA messengers present in the neurons that detect spiciness in the mouth and synthesized a group of corresponding DNA molecules. They then introduced these molecules in various cell cultures to observe the bonding between capsaicin and the proteins produced by the inserted DNA. Finally, they identified the DNA that codes for the receptor of capsaicin, known as VR1. Introducing this DNA sample in frogs' eggs yielded cell cultures whose membrane contained the desired receptor. Further analysis established that VR1 is a membrane channel protein, which regulates the passage of ions (above all calcium ions) between the outside and inside of cells. It was also shown that VR1 has four subunits and that binding with capsaicin opens the channel.

What value do these studies have for gastronomy? Earlier psychophysiological research had led American chemist Wilbur Scoville in 1912 to devise a sort of Richter scale for heat that now bears his name. The electrophysiological recordings of frogs' egg cells equipped with the VR1 receptor demonstrated the soundness of this classification: The cellular response (and therefore, presumably, the neuronal response in the brain) turned out to be proportional to the concentration of capsaicin.

Julius and his colleagues showed that capsaicin, which is fat soluble, can bind itself to the VR1 channel, either on the surface of nerve cells or inside them. Its affinity for fatty substances explains why drinking water does not put out the fire in the mouth, but eating bread does.

Acclimatization to Spices

Exposing the receptors to capsaicin yielded additional information about acclimatization to spicy dishes. The opening of the VR1 channel triggers the inflow of calcium ions into the neuron, which emits a nerve impulse when the intracellular electrical potential reaches a certain threshold. Nonetheless, an upper limit appears to exist as well: Frogs' egg cells equipped with the VR1 receptor die after a few hours of continuous exposure to capsaicin, evidently because the inflow of calcium ions is excessive.

The loss of sensitivity observed in spice lovers seems to result from the death of sensory fibers. This would explain the paradoxically analgesic effect of capsaicin in the treatment of viral and diabetic neuropathies and of rheumatoid arthritis, where by killing pain neurons it helps reduce the sensation of pain.

Finally, the UCSF team showed that rapid increases in temperature trigger ion currents in the VR1 receptor analogous to the currents triggered by capsaicin. The VR1 channel therefore turns out to be both a chemical and a thermal sensor, which is why eating spicy dishes makes the mouth feel as though it is on fire.

28

The Taste of Cold

Cooling and heating the tongue arouse the perception of tastes, even in the absence of food.

THE PHYSIOLOGY OF FLAVOR is riding a wave of fresh discoveries. In recent years physiologists have elucidated the molecular bases of the sensation of spiciness, described the mechanisms in the papillary cells of the mouth responsible for taste perception, and at long last, in April 2000, identified the first receptor for a taste molecule. It was expected that the analytical methods that led to this last discovery would help researchers find other receptors, but a great surprise lay in store: thermal tastes. Variations in the temperature of the tongue alone are enough to cause tastes to be perceived.

When we eat, various molecules stimulate the olfactory cells of the nose, the papillary cells of the mouth, and receptors that register levels of spiciness and a set of mechanical and thermal sensors. How do these different perceptions interact to form the synthetic experience of flavor? It was long supposed that the pieces of information obtained by the various sensory cells traveled upward up from neuron to neuron until they reached the higher centers of the brain, which then gave them a joint interpretation, thus creating the sensation of flavor. A few years ago, however, physiologists came to realize that this account was inadequate: Sensory information has already been combined by the time it reaches the first neuronal relay station, which is to say that sensory integration begins in the tongue.

Heating the Tongue

To determine the physiological consequences of this integration, Ernesto Cruz and Barry Green at Yale University investigated the effect of the temperature of foods on taste perception. Using small thermodes (devices whose temperature is regulated by means of electric currents) to stimulate the tongues of subjects, they discovered—rediscovered, actually—the "thermal tastes" aroused by heating and cooling the tongue. This effect had been discovered in 1964 by G. von Békésy, but the observation was part of a theory that was later discredited and so failed to receive the attention it deserved.

The reactions were not identical for all subjects, but for a large proportion of them, heating the tip of the tongue up to 35°C (95°F) produced a slight sweet sensation, and cooling it down to 5°C (41°F) elicited a sour sensation or, in the case of one person, a salt taste. Heating the back of the tongue produced only a weak sensation of sweetness, but cooling this region often gave rise to more distinct sensations, variously described as bitter or sour. A few subjects tested perceived none of these thermal flavors.

Plainly this curious phenomenon called for further scrutiny. The Yale physiologists sought first to quantify the relationship between temperature and the intensity of sensations: The intensity of the sweet flavor, at the front of the tongue, increases as the temperature rises, and the sour taste caused by cooling becomes saltier as the temperature falls. Because the subjects often reported differences in taste perception between the center and the sides of the tongue, Cruz and Green sought to make these impressions more precise. They found that for all subjects the perception of thermal sweetness was greatest in the tip of the tongue and that thermal sourness was most clearly perceived on the sides of the tongue.

Paradox Elucidated

How are these results to be interpreted? We know that the nerve fibers of two cranial nerves innervate the gustatory papillae. The fibers of the chorda tympani are ramified in the papillae and transmit information about tastes to the brain, whereas the fibers of the lingual branch of the trigeminal nerve terminate in the papillary epithelium, near the taste buds, and transmit information about temperature, pain, irritation, and pressure. Note that thermal sensors and taste receptors are found near one another in the mouth.

It seems plausible to suppose that changes in temperature activate the receptors responsible for the normal coding of the perception of tastes. In this hypothesis, it ought to be possible to block thermal tastes by stimulating taste receptors. Conversely, if the perception of thermal flavors is not caused by the normal taste receptors, inactivating these receptors will have no effect on the perception of thermal taste. Tests are under way.

What are cooks to make of this discovery? One does not taste with the tip of the tongue alone, so the effect of thermal tastes is weak in ordinary eating situations. On the other hand, it is a simple matter to determine whether you are capable of perceiving such flavors: Just place an ice cube against the tip of your tongue or stick your tongue in a glass of warm water.

29

Mastication

Understanding how we chew our food will change how we think about cooking.

DR. JOHN HARVEY KELLOGG, he of the breakfast cereals, advocated a hygienic regime based on relentless mastication. His ideas echoed an ancient East Asian tradition according to which each mouthful of whole-grain rice was to be chewed 100 times.

Why do we chew in the first place? Everyone knows that mastication breaks up food into smaller pieces—small enough that, having also been lubricated by saliva, they easily descend into the digestive system. Jons Prinz and Peter Lucas at the Odontological Museum in London have identified another function. Without knowing it, we chew until particles of food are bound together by saliva into a compact mouthful that can be swallowed in such a way as to minimize the risk that small bits take a wrong turn down into the windpipe. For each food, then, there is an optimal number of masticatory movements.

In asserting that "animals feed, man eats," Brillat-Savarin sought to do away with the animal side of our nature—the very thing that upset the *Précieuses* of mid–seventeenth-century salons in Paris, who made a fashion of mousses because they eliminated the need for "the unsightly act of mastication." And yet who wants to forgo the pleasures of a piece of crusty bread? A sticky dumpling? A crispy piece of bacon? If we are to enjoy the full range of pleasures that the culinary world offers, we must frankly accept our humanity and turn our physiological peculiarities to the advantage of our weakness for good food.

Chewing divides food into pieces of smaller diameter than that of our pharynx. Nonetheless, we normally go well beyond what is necessary for this purpose. As mammals that expend a great deal of energy, we chew our food in order to increase the surface area accessible to digestive enzymes. Indirectly, then, mastication accelerates the assimilation of nourishment.

Prinz and Lucas devised a model to explain how salivation causes the particles formed by chewing to cohere. Their model takes into account the two main forces exerted on masticated food: adhesion between its parts and the adhesion of these parts to the inside of the mouth. These forces depend on the secretion of saliva and the quantity of juice squeezed out of food by the act of chewing.

Small pieces of food are broken up less thoroughly than big pieces. On the other hand, the number of fragments into which a mouthful of food is divided by chewing depends on the mechanical characteristics of the food in question. To simplify the modeling problem, the British physiologists assumed that each piece of food is divided into spherical particles and calculated the total surface forces holding them together.

Furthermore, Prinz and Lucas assumed that these particles agglomerate when the force causing them to adhere to one another is greater than the force causing them to adhere to the wall of the mouth. Using computer calculations of these forces and incorporating values for various other parameters drawn from studies of human physiology, they then determined the cohesion of mouthfuls of food after 150 masticatory cycles for two foods having very different properties: raw carrots, which are broken up very slowly, and Brazil nuts, which are broken up much more rapidly. Computation showed that the cohesion of the masticated food is initially low, then rapidly increases and reaches its highest point after twenty cycles. After that point it diminishes as the particles become smaller and smaller.

To test the proposed model, the calculated degree of cohesion was compared with the cohesion actually measured in mouthfuls of food spit out after having been chewed. The agreement of theory with practice was good, but the actual number of masticatory cycles was a bit higher than the number calculated, no doubt because we are not only machines for absorbing nourishment: "The Creator, in making man eat in order to live," Brillat-Savarin observed, "persuaded him by appetite and rewarded with by pleasure." Because we take pleasure in eating, we prolong our enjoyment by chewing longer than is strictly necessary in order to make food particles cohere.

Model and Cuisine

What can we learn from the model for culinary purposes? Depending on their physical characteristics, foods need a greater or lesser degree of mastication. The addition of compounds that make saliva more liquid (tannins, for example) or increase the concentration of liquids extracted by the teeth has the effect of reducing cohesion, which ought to lengthen the amount of time spent chewing and so add to the enjoyment one takes from a dish. Could this be why gourmets drink wine (which contains tannins) with their meals?

Thickening agents, on the other hand, ought to accelerate the absorption of food into the digestive system. The use of such agents, particularly in diet products, creates a marketing problem: The shorter the time that food is chewed, the fewer the number of odorant and taste molecules that are released.

More generally, the hypothesis that the body automatically detects the ideal cohesion of mouthfuls of food ought to be a source of fresh ideas for the cook who wants to find new ways to combine sticky, gluey, dry, or absorbent ingredients.

30

Tenderness and Juiciness

Chewing is what allows us to enjoy the juiciness and tenderness of meat—though for different periods of time in each case.

LESS THAN A DECADE AGO tenderness was thought to be the most important sign of a good piece of meat. After all, the true gourmet detests tough meat. But how does one tell tenderness and toughness apart? It had been forgotten that meat is not butter and that texture is one of its fundamental qualities. Toughness was confused with a lack of juiciness and the need to chew for a while before swallowing. To elucidate the relationship between the physical structure of meats and their texture, Institut National de la Recherche Agronomique biologists in Clermont-Ferrand analyzed the mastication of samples of meat prepared in various ways.

Studies of the texture of meat have long been hobbled by the mistaken idea that the texture of a food is the same thing as its consistency, which is a microstructural property. Texture has to do instead with the psychological reaction to the physicochemical stimuli aroused by mastication. (For example, water is a liquid, but if you land on it outstretched from a height it can feel as hard as concrete: the texture of water varies depending on whether it can be displaced beneath a falling body, but its consistency is always the same.) Sensory perceptions modulate the motor actions that break up food. Chewing causes the structure of food to be modified, revealing its texture.

Laurence Mioche, Joseph Culioli, Christèle Mathonière, and Eric Dransfield studied the question of texture in the case of meat (in this case beef), searching for similarities between the sensory perceptions aroused by tasting, the mechanical properties of the meat (resistance to compression and cutting), and the electrical activity of the muscles involved in mastication. The beef was prepared in several ways: Some samples were toughened (to a degree that cooking did not subsequently counteract) by immediate cooling after slaughter, and other samples underwent a long aging process at a temperature of 2°C (36°F). Then the different pairs of samples were cooked at 60°C (140°F) and at 80°C (176°F). One piece of each pair was analyzed mechanically, and the other was eaten by trained tasters who judged the elasticity, initial tenderness, overall tenderness, and length of time in the mouth, which is to say the time needed to chew the meat before being able to swallow it. During this exercise the physiologists analyzed the process of mastication by recording the electrical activity of the masseter and temporal muscles.

The Sensation of Toughness

The mechanical measurements corroborated the results of studies that had been conducted for many years at Clermont-Ferrand. Immediate cold storage of food after butchering multiplied by a factor of three or four the resistance to both compression and cutting. Conversely, gradual cooling followed by a prolonged maturation process diminished both types of resistance. Higher cooking temperatures greatly increased the resistance to compression but not to cutting. Finally, differences between the various samples of meat resulted mainly from the action of myofibrillary proteins (responsible for muscle contraction) and the connective tissue, made of collagen, that surrounds the muscle fibers. The physical reactions of the tasters displayed wide variation. The aging period and cooking temperature had perceptible effects on the process of chewing, but differences in preparation had little effect on the electrical activity of the muscles involved.

As they went along the tasters noted their sensations. All of them correctly identified the toughest meats: The type of muscle, the mode of storage, and the cooking temperature affected sensory perception in the same way that they affected mastication. Nonetheless, sensory descriptions did not match the experimenters' predictions. For example, the perception of elasticity did not imply a corresponding initial impression of tenderness. Juiciness, which tasters associated with an initial degree of tenderness, but not elasticity, was influenced more by cooking temperature than by the type of storage; the loss of juice was not perceived as a loss of juiciness. Today we still do not know exactly what juiciness is. Is it the quantity of water in the meat and in the mouth? The quantity of fat? The quantity of saliva secreted in the course of chewing?

The tasters concluded their work by grouping the meats into five classes of increasing tenderness. The meats they found to be the most tender were those that had been aged the longest. The toughest meats were those that had been refrigerated just after slaughter. Lengthening the aging time had a perceptible effect only in the case of meats cooked at 80°C (176°F). Finally, meats cooked at the lower of the two temperatures were thought to be more tender than those cooked at the higher one. Juiciness was found to depend mainly on cooking temperature and much less on the type of storage or aging or on the type of muscle. Differences were plain after the first few bites.

The Reliability of the Senses

Mechanical measurements, sensory evaluations, and electromyographic measurements all yielded the same results, then, with regard to tenderness: Prolonged chewing is needed to make a judgment. By contrast, juiciness is best assessed after a few bites, which detect the general characteristics of the food, causing subsequent mastication to be adapted accordingly.

The Clermont-Ferrand study provided valuable methodological information as well. It revealed that sensory evaluation is the most effective method for detecting differences between various samples. The human perception of the masticatory sequence from beginning to end does a better job of capturing the sensory properties of the meat than mechanical measurements, and the number of masticatory cycles is a more reliable measure of elasticity, tenderness, and toughness, as it is actually experienced in the mouth, than compression measurements. But the mechanical measurement of relative compression is a better guide to juiciness. Tasters are known to adapt their style of chewing to the properties of a particular food, but at which stage of the chewing process they do this merits further study.

31
Measuring Aromas

Chewing slowly deepens the perception of odorant molecules in cooked food.

WHICH AROMAS DO WE PERCEIVE when we eat? For a long time this question could not be answered, for chemical analysis was unable by itself to determine the concentrations of odorant molecules in the vicinity of the receptor cells in the nose. Andrew Taylor, Rob Linforth, and their colleagues at the University of Nottingham, working in association with Firmenich (an international perfume and flavor research group), have been conducting experiments since 1996 with a device that shows how aromatic compounds are released during the mastication of food. The same food, it turns out, smells different to different people.

Odorants—volatile molecules that stimulate the nasal receptors in passing upward from the mouth through the rear nasal fossae as food is chewed—are important components of flavor. Nonetheless, their sensory action is difficult to analyze because these molecules interact with saliva and with various other compounds present in foods. Accordingly, the odorant profile of a particular food cannot be reduced to its chemical composition.

Because molecules can be detected by smell only if they pass into the vapor phrase, physiologists have sought to measure the concentration of odorants in the air above foods. But given that the chewing of food, breathing, and salivation all affect the release of aromas, one cannot rely on this measurement alone. To identify the active aromas of a food, it is necessary also to measure the release of odorant molecules while it is being consumed.

The new method of mass spectrometry devised by Taylor and his colleagues directly measures the concentrations of odorant molecules in the breath of subjects as they are chewing food. A stream of gas containing the volatile molecules to be analyzed is pumped into a chamber equipped with an electrically charged needle that ionizes water molecules. The hydrogen ions that are formed in this way then transmit their electrical charge to the odorant molecules, which are attracted by a series of electrically charged plates in a focalization chamber. From here they are channeled into another chamber for analysis.

The Wisdom of Chewing Slowly

The British chemists first examined how a gel composed of gelatin and saccharose releases the volatile components—ethylbutyrate, found in fruits such as strawberries, and ethanol—that are trapped in it. A tube was placed in a nostril of each of the subjects (who were nonetheless able to breathe without difficulty) to capture a sample of the air present in the naval cavity.

The first observation was not surprising. Because molecular concentrations in the air in the nose vary periodically with the rhythm of respiration, what one wants to know, for each breath taken and expelled, is the maximum measured concentration. In the case of acetone, the maximum concentration was the same for each respiratory cycle because this molecule, which is released by the metabolism of fatty acids in the liver, is naturally found in the breath.

By contrast, the ethylbutyrate and ethanol detected in the breath came from the gel alone: The ethylbutyrate was released only during mastication, for about a minute, whereas the ethanol was released for a longer period of time. Because ethanol is soluble in water, it dissolves in saliva after having been released by the rupture of the gel, and only afterward does a part of it pass into the air. This slow exchange between water and air is stimulated by chewing and continues even after chewing is finished.

These studies also confirmed what the makers of chewing gum have long suspected, namely that the release of odorant molecules depends on both the speed of mastication and the softness of the gum. A comparison of the reactions of three people to the same food (a gel made from gelatin containing ethanol, butanol, and hexanol) revealed what might be called odorant inequality: The maximum concentration of molecules in the breath and the time it took for this concentration to appear varied according to the rate of mastication for each person tested. The maximum odorant concentrations were lowest in the case of the most rapid eaters, presumably because they broke up the gel the least. Brillat-Savarin therefore was right to say, "Men who eat quickly and without thought do not perceive the [succession of] taste impressions, which are the exclusive perquisite of a small number of the chosen few; and it is by means of these impressions that gastronomers can classify, in the order of their excellence, the various substances submitted to their approval" (Meditation 2, *The Physiology of Taste*).

32

At Table in the Nursery

Observing the eating habits of small children provides clues to their developing appreciation of food.

FOOD-MINDED PARENTS ARE FOREVER COMPLAINING that their children like only starches (pasta, rice, potatoes), the blandest cheeses, and tasteless meats (especially the white meat of chicken). Why do we begin our eating lives liking such dull foods? How can children be set straight about food before it's too late? It used to be that psychologists and sociologists were likeliest to wonder about the biological motivations that cause parents to despair for their offspring. Today it is the sensory biochemists who have taken the lead in exploring the dietary preferences of children and how these change.

An important experiment that only now is yielding its first results began in 1982 and finished in 1999 at the Gaffarell Nursery of the Dijon Hospital in France. The children, aged two to three years, were allowed to decide what they would eat for lunch, with several constraints. The menu contained eight items: bread, two starters, meat (or a meat-based dish) or fish as a main course, two vegetables or starches, and two cheeses. Sweets were not offered as part of the meal because of the children's presumed attraction to sugar; they were reserved for snacks instead. The children could have extra helpings but not more than three of the same food during the same meal. Their minders accommodated and recorded the choices freely made by the twenty-five children enrolled in the nursery each year. All eight dishes were placed on the table, and the children were free to serve themselves as they pleased or to eat nothing at all. In all 420 children were monitored, each for an average of 110 meals.

The results are being scrutinized by Sophie Nicklaus and Sylvie Issanchou at the Laboratoire de Recherches sur les Arômes at the Institut National de la Recherche Agronomique station in Dijon, in collaboration with Vincent Boggio of the medical school at the University of Dijon. They are interested in a number of questions. What foods did the children eat? Did they show a general aversion to certain dishes? How are different choices between dishes to be interpreted? How are variations in choice to be explained: by individual preferences? Sifting through the data, the researchers found evidence in favor of certain familiar assumptions as well as a number of unanticipated outcomes.

A Taste for the Tasteless

First, as parents have long maintained, children do indeed prefer starches and meats on the whole. The cheeses they select are almost invariably ones with minimal taste and a soft texture. Rare are the *amateurs de roquefort* at the age of two or three—gourmets in short pants one day perhaps, but not in diapers. It may be a sign of the times that bread was the starch least often selected, behind pasta, rice, and fried potatoes. The most popular dishes were French fries, small sausages of the kind known in France and elsewhere in Europe as chipolatas, quiches, pasta, breaded fish, rice, mashed potatoes, ham, and beefsteak. No surprise, any of this—only a quantified confirmation of what has long been observed.

On the other hand, the choice of meats and vegetables did hold surprises for the researchers. The children made little distinction between them, choosing roast pork, turkey, leg of lamb, and organ meats almost indiscriminately. But when it came to vegetables, which were chosen less often than meats, preferences were quite clear, depending on the type of vegetable and the type of culinary preparation. Spinach, which is so widely thought to be disliked by children, was selected more often and consumed in greater quantities than any other vegetable—as long as it was napped with a white sauce. Endives, cabbage (raw or cooked), tomatoes, and green beans appealed to few of the young diners, however. It appears that foods with a hard and fibrous texture (which makes them hard to chew) are more likely to be refused, as are those with a pronounced bitter flavor.

How are we to account for all this? It occurred to the sensory biochemists in Dijon to try to relate the selection of foods to their nutritional content. What they discovered was that the higher the caloric value of a food, the more often it is chosen (with the exception of cheeses). Is this proof that innate dispositions are still strong in young children? We know that infants who are fed salt, sweet, sour, or bitter liquids react visibly with disgust in the case of the last two flavors but show pleasure when they encounter a sweet taste. In this they resemble fruit-eating monkeys, who associate sweetness with the presence of sugars (whose molecules have a high energy value) and bitter with vegetables that contain toxic alkaloids. Gradually children come to learn, through conditioning and culture, to diversify their diet. Conditioning causes children to associate satiety, for example, with the consumption of high-calorie nonsweet dishes (fatty foods in particular); culture accounts for the fact that some children, even when quite young, learn to relish dishes with a powerful taste, contrary to what purely biological explanations would predict.

All Equal, All Different

None of the factors analyzed so far explains the highly variable choice of foods with a strong taste—neither sex, nor maternal feeding, nor the child's birth order, nor his or her body mass index (a measure of fatty mass obtained by dividing weight by the square of the child's height).

Because the dietary preferences of young children have been little studied until now, it is not surprising that these first studies have left many observations unexplained. Moreover, it will be a long time before we are able to correlate the choices of the children studied with their preferences as adults. But patience is the price we must pay if we are to identify the mechanisms of dietary learning in childhood and to clarify the role of early exposure to different foods in the formation of later preferences—long thought to be important but so far undemonstrated. The food industry takes an obvious interest in the results of these studies, for it will be able to offer products of longer-lasting appeal only if the dynamics of the dietary preferences of young children are understood.

33
Food Allergies

Predicting and preventing the risks of allergic reaction to transgenic foods.

GASTRONOMY IS NOT A WORLD ONLY of pleasurable aromas and tastes. More than one-quarter of the population in seven countries of the European Union claims to suffer from food allergies or intolerances. Although clinical tests indicate that the actual incidence of such allergies is much lower than commonly believed (about 3.5% of the population), these reactions pose a major public health problem, all the more because the seriousness of the attacks reported is on the rise. In the last ten years the number of cases of anaphylactic shock caused by a food allergy has quintupled, and many of these attacks are fatal, notably ones triggered by the ingestion of products derived from peanuts.

Jean-Michel Wal and his colleagues at the Institut National de la Recherche Agronomique-Commissariat de l'Énergie Atomique (INRA-CEA) Laboratoire d'Immuno-Allergie Alimentaire in Saclay have shown how a protein found in cow's milk triggers allergies despite its resemblance to a human protein that is readily tolerated. They are also conducting animal tests of a method of gene immunization for preventing allergies.

The reasons for the recent growth of food allergies are not completely understood. The increase in the number of offending foods seems to be linked to the development of respiratory allergies to plant pollens; the antibodies produced by the human organism against a pollen antigen sometimes react also against molecules from a very different source. These cross-reactions are a product of shared epitopes, molecular fragments that are the target of the immune system. Thus the multiplication of allergies to exotic fruits (avocados, kiwi fruits, bananas, and so on) has been found to be associated with the development of allergies to latex, themselves on the rise as well.

To these traditional foods have been added new ones derived from genetically modified organisms, which contain proteins normally absent in traditional foods. The growing commercial availability of such products has aroused concerns that they may be allergenic, but because few cases of illness have been reported until now it is not clear how to assess the risk posed by the introduction of new foods. Several approaches are being investigated: tests involving serums from allergic patients (whose antibodies bind to allergenic molecules), cutaneous tests (in which molecules thought to be allergenic are deposited on the skin, and signs of adverse reaction are monitored), animal studies (in which candidate molecules are administered to sensitized animals), and theoretical models of the incidence of allergic reactions based on the protein sequence in amino acids.

Developing such models will take time, but the preliminary step of compiling sets of well-characterized data for proteins whose allergenic character is known is now being carried out. Wal and his colleagues have already used these databases to compare the immune reactivity of human and bovine beta-casein.

Allergies to Milk

Caseins constitute 80% of the proteins found in milk. There are four types: alpha S1, alpha S2, beta, and kappa. The beta-casein in cow's milk is an important allergen: 92% of the serums obtained from people allergic to caseins contain a type of antibody responsible for allergic reactions, known as immunoglobulin E, that is specifically directed against beta-casein.

In 1997 the immunologists at Saclay demonstrated the presence of immunoglobulin E molecules specific to a bovine whey protein in the blood of allergic people and observed that these molecules also react with the corresponding human protein. Does this same kind of cross-reaction exist for beta-casein? Wal and his colleagues tested the serum of twenty people allergic to the beta-casein of cow's milk, carefully measuring the concentration of immunoglobulin

E specific to bovine beta-casein and then verifying that these molecules also reacted with human beta-casein.

Human and bovine beta-caseins share 50% of their amino acids, hence the interest in studying their common domains. One such domain corresponds to an amino acid sequence that assumes a helical form in both human and bovine proteins. Another common domain contains the principal phosphorylation sites of beta-casein. Once the amino acids that make up this protein have been assembled, it is modified in particular by the addition of phosphate groups, a process known as phosphorylation. Such posttranslational chemical modifications influence the properties of the molecules by changing their electrical charge and altering their conformation. The INRA-CEA immunologists were able to show that the phosphorylation of this second domain affects the allergenic character of beta-casein.

Preventive Measures

Preliminary studies of new foods derived from genetically modified organisms have not yet shown any of them to cause allergies. However, research is being conducted on possible preventive measures. In the case of beta-casein, milk substitution has proved to be ineffective because the caseins of different kinds of milk are very similar. The Saclay team therefore turned its attention to testing gene immunization in mice. This involves injecting the animals with bacterial DNA containing additional genetic material that codes for a particular protein. These sequences activate a nonallergic reaction of the immune system that switches off the allergic reaction. In the case of allergies to bovine beta-lactoglobulin (another protein found in cow's milk), the method was found to substantially and persistently reduce the production of immunoglobulin E specific to this protein.

34
Public Health Alerts

Food warnings attest to the progress being made by microbiology.

CASES OF LISTERIOSIS ARE DRAMATIC, sometimes fatal, and all the more shocking because measures to protect public health have never been more effective. Nonetheless this relative security cannot be counted on to prevent the occurrence of illness, even death, among people in the leading risk groups: pregnant women, the elderly, young children, and those with immunodeficiencies. Because health alerts are issued as a consequence of retrospective epidemiological inquiries that compare strains isolated from the victims with products suspected of contamination, they constitute a response to crisis rather than a means of preventing it. One of the profound challenges facing basic research in microbiology is to develop a better understanding of *Listeria* and other food microorganisms.

The most urgent task is to identify pathogenic strains. Only when these are known can the populations at risk be determined and a range of commercially available foods tested. At the Institut National de la Recherche Agronomique Laboratoire de Pathologie Infectieuse et Immunologie at Nouzilly, Patrick Pardon and his colleagues are studying the comparative virulence of *Listeria*. The food-related strains of this bacterium that have been isolated have variable pathogenicities. How can dangerous strains be distinguished from benign ones? In recent years microbiologists have distinguished two types: *Listeria monocytogenes*, considered always to be pathogenic, and a harmless variety called *Listeria innocua*. But even this classification, which has made it possible to avoid issuing unnecessary alerts, is inadequate. The problem is that present-day techniques do not allow the virulence of different strains to be characterized with precision.

Rapid Testing for Necessary Alerts

The virulence of different strains has long been studied in laboratory animals (chiefly mice) inoculated with microorganisms. This method is both time consuming and costly, and it should be abandoned once alternative procedures become available.

The Nouzilly microbiologists developed an in vitro test that uses human intestinal cells, which offer a point of entry to the organism. To test the pathogenicity of a particular strain of *Listeria,* one places a sample in suspension on a culture of these cells. One then looks to see whether ruptures appear in the cell layer in response to the introduction of the bacteria.

By means of this test the biologists were able to discover that the *Listeria monocytogenes* group is not homogeneous: Certain strains were virulent to only a small degree or not at all (the results were confirmed on laboratory animals), which led to further analysis of the genome and the proteins of these bacteria. A three-year national program is under way in France in which several teams of researchers are reviewing a database of about 400 strains of *Listeria monocytogenes.* The genes and proteins of strains that pose only a slight pathogenic risk, or none at all, will then be compared with those of virulent strains in the hope of further refining the detection of pathogenic strains.

In parallel with this work, research is being conducted with a view to identifying food products and processing techniques that are conducive to the development of pathogenic strains. This means determining the levels of humidity, temperature, acidity, and storage time associated with a risk of contamination in order to improve industrial quality control (the legal standards for which vary according to the product) and official supervision by government agencies.

The Risks of Eating

The dissemination of food contamination information is bound to be unsettling for producers and consumers alike. But it is necessary because it prevents illness by forcing producers to take contaminated foods off the market. The fact remains that certain foods pose a greater danger to public health than others. Must they be banned? In Patrick Pardon's view, there is no point trying to prohibit the sale of products that carry a risk of *Listeria* contamination; France will not consent to being deprived of its rillettes or raw milk cheeses. The duty of public policy is to minimize the spread of infection by warning those who are most vulnerable to it. Prohibition would have adverse consequences not only for the economic stability of the food processing industry but also for the health and well-being of the public, which would be tempted to provide itself with banned products from unsafe sources.

3 Investigations & Models

PART THREE

35

The Secret of Bread

Chemists look to improve bread dough by investigating the protein bonds that form its glutenous network.

THE BEHAVIOR OF WHEAT FLOUR can be understood by analyzing the properties of its two main components: starch granules, which swell up in the presence of water, and proteins, which form a glutenous network as dough is kneaded. How do the forces among proteins contribute to the mechanical properties of the dough? It has long been known that bonds between the sulfur atoms found in wheat proteins play a role in structuring gluten. Other forces have been discovered as well.

Gluten is a viscoelastic network of proteins that becomes elongated by pulling and then partially reverts to its initial form when the tension is relaxed. The quality of bread depends on the quality of its gluten. Indeed, gluten is what makes breadmaking possible: Yeast produces carbon dioxide bubbles, with the result that the volume of the dough increases, and the protein network of the gluten preserves the dough's spherical shape by retaining these gas bubbles. It therefore becomes necessary to understand the reticulation of wheat proteins, that is, how bonds are established between them.

As early as 1745 the Italian chemist Jacopo Becarri showed that gluten can be extracted by kneading flour with a bit of water and then placing the lump formed in this way under a thin stream of water. Rinsing washes away the white starch granules, leaving the gluten between one's fingers. It was later demonstrated that only certain insoluble wheat proteins called prolamins make up the glutenous network of bread.

These prolamins are of two types: gliadins, which are composed of only a single protein chain (a sequence of amino acids), and glutenins, which are large structures composed of several protein subunits linked by disulfide bridges (that is, the subunits are connected by two covalently bound sulfur atoms). Do these disulfide bridges also link the glutenins to one another? The traditional view is that kneading establishes supplementary disulfide bridges between the various prolamins that break and almost immediately reform as the baker works the dough.

The glutenins have a central domain (containing 440–680 amino acids) formed of short repeated sequences and flanked by two terminal domains. The size of the central domain determines the molecular mass of the glutenins; the terminal domains contain cysteines, amino acids that bear sulfur atoms capable of forming disulfide bridges. Nonetheless, the chemical characteristics of glutenins do not completely explain their capacity to make gluten.

A World Made of Dough

In 1998, a team led by Jacques Guégen at the Institut National de la Recherche Agronomique station in Nantes showed that prolamins can bond with one another by means of dityrosine bonds. Tyrosine is an amino acid whose lateral chain is composed of a CH_2 group, a benzene nucleus, and an –OH hydroxyl group. Shortly afterward, on the basis of this research, Katherine Tilley and her colleagues at the University of Kansas demonstrated the importance of dityrosine bonds in gluten. From bread dough at various stages of kneading, they extracted, dissociated, and chemically analyzed the gluten of the kneaded flour and found that concentrations of dityrosine increased during kneading. This raised the question of what role dityrosine plays in the formation of gluten.

Further analysis disclosed the existence of two types of dityrosine bonds: dityrosine, in which two benzene groups are linked by a neighboring carbon atom of the –OH hydroxyl group; and isodityrosine, in which an oxygen atom belonging to a hydroxyl group on one tyrosine binds to its carbon neighbor in the hydroxyl group on the other tyrosine.

This discovery caused a stir among gluten chemists, for dityrosine bonds are commonly found in plant proteins, whose sequences and structures resemble those of glutens, as well as in resilin proteins, found in insects and arthropods, and in elastin and collagen, both found in vertebrates. In forming dityrosine bonds by kneading dough, the baker reproduces the living world.

On the other hand, the Nantes team observed that dityrosine bonds occur in the presence of a type of enzyme known as peroxidase, which is naturally present in flour. Does the long working of the dough needed to make bread cause the enzymes to react with the glutenins by giving them time to establish dityrosine bonds? What roles do dityrosines and disulfide bridges play in the formation of gluten?

The ability to identify these bonds raises a further question. Improved additives can be used to facilitate kneading or intensify the production of gluten. When oxidant compounds such as ascorbic acid and potassium bromate are added to bread dough, for example, the number of dityrosine bonds that are formed increases. It used to be thought that additives of this sort favor the formation of disulfide bridges, but it may be that they also cause dityrosine bonds to be created. In that case one may imagine new methods for selecting wheat on the basis of its gluten content. Would it be enough simply to measure the quantity of dityrosines in a certain kind of dough in order to assess the quality of its gluten?

36

Yeast and Bread

Bread owes its flavor to fermentation.

FRENCH BREAD—particularly its principal representative, the baguette—is reproached nowadays for having less flavor than it used to and for drying out too quickly. The second criticism is unjust, for it neglects the fact that the baguette was a product meant for city dwellers who could buy what they needed several times a day at their local bakery; the crust was more important than shelf life. Yet many bakers today admit to being more concerned with the mechanical behavior of the dough than the taste and odor of the bread.

The flavor of bread comes from the cooking of the dough, in which various reactions combine to produce a crispy, flavorful crust, and from the action of the constituent molecules of a species of yeast known as *Saccharomyces cerevisiae* (brewer's yeast) and the products of their fermentation in helping to form the volatile compounds found both in dough and in bread itself. To better understand this process, two Institut National de la Recherche Agronomique (INRA) groups in Nantes compared breads made by several different methods, with or without yeast and with or without fermentation.

The most common method, direct yeast fermentation, begins with the kneading of a dough composed of flour, water, yeast, and salt for about twenty minutes. The dough is then allowed to ferment for forty-five minutes (floor-time), at which point it is divided into lumps and fermented again for an hour and forty minutes (proofing) before being cooked at 250°C (about 475°F) for thirty minutes.

The sponge method (sometimes called *sur'poolish* or French starter) is identical except that the dough undergoes a process of prefermentation in a semi-liquid environment. Water is first combined with a smaller quantity of flour to obtain a preparation with the consistency of crêpe or pancake batter. This preparation is left to ferment for several hours. An additional measure of flour is then incorporated to restore a consistency equal to that of directly fermented dough, and the same procedure followed as with the traditional method.

A third method consists of cultivating beforehand a natural microflora composed of yeast and lactic bacteria. The starter thereby obtained—sourdough—is then used to initiate the process of fermentation in bread dough.

"Vinegar" Needed

As late as the mid-1980s studies of the volatile compounds in bread revealed no systematic qualitative differences from one type of bread to another, although tasters were quite capable of distinguishing between them. Only the proportion of certain volatile organic acids seemed to vary markedly. In particular, the sponge and sourdough methods yield acetic acid concentrations that are, respectively, two and twenty times greater than those obtained by direct fermentation. In the case of the sourdough method, lactic acid is also produced by the bacteria that colonize the dough along with the yeast.

Studies of direct fermentation extended these first researches by clarifying the action of yeast in bread, a complex environment that is modified by cooking. An initial comparison of breads obtained by direct fermentation with breads baked after inhibition of yeast, without the addition of yeast, or with yeast added just before cooking revealed the various compounds produced by the action of yeast in French breads. The use of yeast in fermenting dough is directly responsible for the presence of compounds such as 3-hydroxy 2-butanone, 3-methyl 1-butanol, and 2-phenylethanol (which has the odor of a wilted rose).

In bread cooked without yeast, other components are more abundant than in ordinary bread, particularly monounsaturated and polyunsaturated aldehydes and alcohols such as pentanol and benzyl alcohol, which may result from the oxidation of lipids in flour (the analogue of foods that turn rancid), a process known to contribute to the flavor of bread. The constituent compounds of yeast appear to play only a weak role.

Finally, the complexity of the phenomena that produce the transformation of bread dough led the INRA biopolymers laboratory in Nantes to study uncooked dough as well, with or without yeast, fermented or not. In this case chemical analysis was combined with investigation of the olfactory properties of dough extracts: Volatile compounds were extracted by various solvents, separated by chromatography, and smelled by the experimenters through a tube.

What they discovered was a general increase in the concentration for several alcohols, ketones, esters, and lactones and a reduction in the concentration of aldehydes. They also found that yeasts produce higher levels of alcohol. But chromatographic analysis, combined with sensory evaluations, showed that these levels have little to do with the aroma of the fermented dough, unlike the aldehydes and two compounds that so far have not been identified.

Such studies are being published at a time when bean and soy flours, customarily added to bread as whitening agents, are used less and less often. When the dough is subjected to longer and more rapid kneading by mechanical means, these flours produce hexanol, which brings a stale, oily smell. In reviving neglected techniques of fermentation such as the sponge method, a new generation of bakers seeks to avoid the disadvantages of standardized production and to reinvigorate the market for a food that is as old as humanity itself.

37
Curious Yellow

The unsuspected nature of an egg yolk.

BEHOLD THE EGG, WHOLE AND RAW, in its shell. Where is the yolk? No doubt, the physicists will say, on a vertical axis—for reasons of symmetry. Yet there are other possibilities. The yolk could be in the upper part, in the center, or in the lower part. How can we determine its location? Put a yolk in a tall, narrow glass and cover it with several whites: The yolk rises to the top, which suggests that the same thing occurs in a whole egg.

But could it be that the membranes that surround the yolk and bind it to the rest of the egg prevent the yolk from rising in its shell? There are several ways to show that this is not the case. If you boil an egg standing on its end and examine the yolk you will see that it is lodged in the upper part of the shell. Coagulation may have disturbed the internal arrangement of the egg, you say? Place a whole fresh egg in its shell in vinegar. After two days or so, the shell will be dissolved by the vinegar, and you will see the yolk floating on top of the white (the egg retains its shape because it is held in place by membranes and because the external layers have been coagulated by the action of the vinegar). Or try an even simpler experiment: Remove the top of the shell of a fresh egg and look to see where the yolk is.

More elaborate procedures can be followed as well. Although radiography gives poor results (because the shell is opaque to X-rays), ultrasound yields surprising results. In images obtained by immersing an ultrasound probe in an egg through a hole in the top of the shell, the yolk can be seen to be composed of concentric layers similar to those of a tree.

Why is it that no one who has eaten a soft-boiled egg has ever suspected the existence of this structure? The yolk is an alternation of layers called deep yellow, 2 millimeters thick, and clear yellow, 0.25–0.4 millimeters thick. These layers are produced by the hen during the day and during the night, respectively. The difference between the two types results from the rhythm of feeding, which produces a weaker concentration of yellow pigments during the night than during the day. Of course, these layers become mixed together when you pierce a yolk, so you cannot see them.

Granules and Plasma

If we continue our investigation under the microscope, we see that the two layers, light and deep yellow, are not homogeneous. Instead they are composed of granules dispersed in a continuous phrase called plasma. Marc Anton and his colleagues at the Institut National de la Recherche Agronomique station in Nantes separated the granules from the plasma by centrifugation and observed that roughly half of the yolk is made up by water, a third by lipids, and about 15% by proteins. Proteins and lipids often are associated in particles that are distinguished according to their density: low-density lipoproteins (LDLS) in the plasma and high-density lipoproteins (HDLS) in the granules. Isolating them makes it possible to test their properties. For example, it can be shown that the LDLS combine to form a gel when they are heated to a temperature of about 70°C (158°F). It is these structures—composed of proteins and lipids (notably cholesterol)—that are responsible for the setting of the yolk during cooking.

It has long been claimed that mayonnaise, which consists of droplets of oil dispersed in water (from either egg yolks or vinegar), is stabilized by lecithins and other phospholipids in the yolk. Anton and his colleagues sought to answer this question by determining whether the emulsifying properties of the yolk come from the plasma or the granules. Because the solubility of proteins depends on acidity, the Nantes biochemists began by studying their solubility in terms of pH (a measure of acidity) and salt concentration. They found that plasma proteins are completely soluble at all levels of pH and all degrees of salt concentration, whereas the solubility of the granular proteins varies: They have low solubility at a pH of 3—that of mayonnaise—but become more soluble at neutral pH in a low-salt environment (sodium ions replace calcium ions, which establish bridges between the granular proteins inside the granules, with the result that these proteins are released).

Protein solubility is not the only thing that must be taken into account in order to make a successful emulsion, however. The less upward movement there is by the oil droplets in the water phase, the more stable the emulsion. In the plasma this movement is minimal for a pH of 3, and salt concentration has no effect in an acid environment; emulsions obtained from granules, on the other hand, are sensitive to both acidity and salt concentration. Emulsions made with whole egg yolks behave like those obtained with plasma.

In sum, the component elements of plasma are responsible for the egg yolk's emulsifying effect, and proteins do a better job than phospholipids of preventing the oil droplets from moving upward, thereby stabilizing the emulsion. Is this because the proteins in the LDLS of plasma act by electrostatic repulsion at the surface of the oil droplets, causing the droplets to repel one another? Or because they protrude from the surface of the droplets and act instead by steric repulsion? At a pH of 3, proteins are electrically charged and repel one another; at a pH of 7, however, it is the proteins' steric properties that stabilize the droplets by blocking their tendency to fuse with one another. The exact mechanisms of this behavior have yet to be understood.

All this for an ordinary egg yolk.

38

Gustatory Paradoxes

The environment of aromas affects our perception of them.

A GLASS OF VINEGAR IS UNDRINKABLE, but it becomes palatable if one adds a large amount of sugar to it. Yet the pH—the acidity of the vinegar measured in terms of the concentration of hydrogen atoms—is unchanged. Why is the sensation of acidity weakened? Because the perception of tastes depends on the environment in which taste receptors operate. The same interactions take place in the vicinity of the olfactory receptors. Chantal Castelain and her colleagues at the Institut National de la Recherche Agronomique station in Nantes sought to identify these synergies in order to lay the basis for a rational theory of the aromatization of foods.

Most foods consist of water and fats, which are insoluble in water. When foods are put in the mouth, the taste molecules reach the taste receptors after having first diffused through the saliva (a watery solution). Simultaneously the odorant molecules pass into a vapor phase, in the air contained in the mouth, and from there migrate toward the olfactory receptors in the nose. This simple description is complicated by the fact that taste and odorant molecules are never completely soluble or insoluble in water and that, depending on their chemical composition, they are variably distributed in liquids and air. It is clear, then, that the aromatization of foods can be mastered only if the diffusion of molecules between liquid phases (either water or oil) and gaseous phases is understood.

To analyze these various migrations, the Nantes biochemists constructed a device in which a neutral oil can be placed in contact with water and one can measure the distribution of molecules among four compartments: water, oil, the air above the water, and the air above the oil. They began by dissolving the molecules, in either the water or the oil, and then measured their concentration in the three other compartments. The aromatic molecules were found to follow different trajectories depending on experimental conditions. When dissolved initially in oil, for example, they pass into the air above the oil while diffusing in the water and passing from there into the air above.

Several molecular mixtures were tested: esters, aldehydes, alcohols, and ketones. Paradoxically, transfer was observed to be more rapid from oil to water than from water to oil in the case of the esters and ketones but more rapid from water to oil than from oil to water in the case of the alcohols and aldehydes. Because many alcohols are soluble in water, for example, one would have expected them to have trouble migrating toward oil, in which case the transfer ought to be slower from water to oil than from oil to water.

In the Nose

To better understand how foods are aromatized, these studies must be supplemented by an analysis of the effects of odorant molecules on the receptors in the nose once they have penetrated the mucus layer covering these receptors and dissolved in the hydrophobic phases of the cellular membranes. The Nantes researchers are investigating the perception of molecules above the water and oil phases. Having dissolved a single molecule in two liquids—oil and water—in concentrations such that a sensor registers the same partial pressure of the molecule above each liquid, they analyzed the perceptions of people who breathed the air above the two.

Paradoxically, again, the human nose sometimes perceives a difference. For certain molecules, the nose registers a very similar result to that of the detectors (in the case of linalol, for example, which has an odor of lavender and bergamot, depending on the concentration), but no general law has been deduced because other odorants such as 1-octane 3-ol (which has a smell of mossy decomposing undergrowth and mushroom), benzaldehyde (a smell of almond), and acetophenone (a smell of beeswax) give rise to different perceptions. Obviously the presence of water vapor can affect the perception of the aroma.

Such phenomena are likely to be found in connection with taste as well. The texture of foods determines flavors by affecting the length of time during which taste and odorant molecules remain in the mouth and the rates of diffusion of these molecules from foods toward the olfactory and gustatory receptors. Thus a mayonnaise that is too acid becomes less so when it is beaten because its firmer texture slows down these transfers.

39
The Taste of Food

The texture of vinaigrettes determines their odor.

COOKS WELL KNOW THAT ADDING too much flour to a sauce makes it tasteless. The reason is that the flavor of foods does not depend solely on the odorant and taste molecules they contain but also on interactions between these molecules. Molecules with no odor, such as proteins and starch molecules, bind with certain odorant molecules and prevent them from acting on our senses. One would like to know exactly which aromas are masked in this way.

Identifying all the chemical bonds between the various molecules in a food is not enough, for its physical structure plays a role as well. Even in homogeneous phases such as solutions, the release of odorant molecules depends on the viscosity of the system. However, foods are not solutions but dispersed systems. In foams, air bubbles are trapped in liquids or solids; in emulsions, oil droplets are dispersed in water; in suspensions, solid particles are distributed in liquids; and so on. The odorant molecules that are present within the dispersed or continuous phases of such physicochemical systems are not released in the same fashion as if they are simply dissolved.

Complicating matters still further is the fact that in emulsions, for example, dispersed droplets of oil are held in place by layers of tensioactive molecules. Because of their solubility in both oil and water, these molecules almost necessarily bind with odorants, which then become inaccessible to the olfactory system. To identify the phenomena that determine the release of odorant molecules, Marielle Charles, Élisabeth Guichard, and their colleagues at the Institut National de la Recherche Agronomique and École Nationale Supérieure de Biologie Appliquée à la Nutrition et l'Alimentation stations in Dijon studied salad dressings in collaboration with researchers at Amora-Maille, the well-known producer of mustards and condiments.

Fragrant Salad Dressings

In the vinegar-based sauces studied, the watery phase of these emulsions consisted of a mixture of wine vinegar, lemon juice, and salt. Drops of sunflower oil were then emulsified in it (that is, kept in a dispersed state) thanks to the action of whey proteins. Finally, a mixture of xanthan (a polymer obtained by microbial fermentation of glucose) and starch was incorporated to stabilize the sauce. To this basic salad dressing the physical chemists added fixed quantities of odorant molecules: allyl isothiocyanate, in the oil phase, which yielded a hint of mustard; and, in the water phase, phenyl-2-ethanol and ethyl hexanoate, which gave rose and fruit notes, respectively. When several emulsions were compared each one was found to contain oil droplets of a particular size. A jury of trained tasters evaluated the sauces, noting the intensity of the various accents—lemon, vinegar, mustard, and so on.

The results of these taste tests were difficult to analyze: Although the acid taste was preponderant, the tasters struggled to describe the other sensations. Their training nonetheless enabled them to perceive that the intensity of the overall odor, as well as that of the taste and odor of egg, the mustard odor, and the butter taste, increased with the size of the oil droplets and, conversely, that the intensity of the citrus fruit odors diminished as the oil droplets got larger.

Migrations of Odorant Molecules

To interpret these observations more precisely, the researchers analyzed the concentrations of volatile molecules in the air above the sauces. They found that the water-soluble components were present in lower concentrations as the oil droplets became smaller and, conversely, that in this case the oil-soluble molecules were more abundant.

What accounts for these differences? Consider first the oil-soluble odorant molecules. When emulsions containing fixed proportions of water and oil are vigorously whisked, the oil droplets become smaller and more numerous. The lipophilic molecules therefore have a shorter distance to cover in order to reach the surface of the droplets. Moreover, because the total surface area of the oil–water interface has increased, the tensioactive molecules form a thinner layer above the surface of the oil droplets, with the result that the lipophilic odorant molecules diffuse more easily outside the droplets. Finally, the extent of the water layer (in which these odorant molecules bind with the thickening agents) that must be traversed before they reach the air above the emulsion is smaller.

For the odorant molecules that are more soluble in water, the effect is opposite: The smaller the oil droplets, the harder it is for these molecules to diffuse in a thinner layer of water, and so they are less readily perceived.

Are these general laws? Because foods contain various kinds of molecules, many studies will be necessary to characterize all the relevant mechanisms and then to elucidate their relative importance. "Life is short," said Hippocrates, "the art is long, opportunity fleeting, experience misleading, judgment difficult."

40

Lumps and Strings

The enemies of successful sauces form because water diffuses slowly in starch paste and gelatin.

LUMPS—THE SHAME AND DESPAIR of the cook. Cookbooks suggest various ways to avoid them in making béchamel and other sauces that are thickened with flour. Some authorities say to make a roux first, cooking butter and flour and then adding milk (cold according to some, hot according to others); others insist on the opposite method, namely pouring the roux into the milk, which, again depending on the author, should be either cold or hot.

Which method is better? If we are willing to waste a bit of flour, butter, and milk in order to test the four possibilities we will find that the formation of lumps is at bottom a question of speed: When the roux is added gradually to the milk or the milk to the roux, lumps do not appear; however, they do form when the two ingredients are mixed together at once, especially when the roux is poured into boiling milk.

This experiment gives us a method, but it does not explain the phenomenon. Let's study the matter further by simplifying it. The butter serves chiefly to cook the flour, eliminating its bland taste by creating odorant and taste molecules (the browning is associated with the chemical reactions that form these molecules), but it is not the cause of the lumpiness, which also results from combining flour and water. Why? Flour is composed principally of starch granules, themselves made up of amylose and amylopectin. These two types of molecule are both polymers, which is to say chains of glucose molecules (linear in the case of amylose, ramified in the case of amylopectin).

Amylopectin is insoluble in water, but amylose is soluble in hot water. When the starch granules are put into hot water, they lose their amylose molecules, and the water fills the space between the amylopectin molecules, causing the granules to swell up and form a gel known as starch paste.

This description gives us a handle on the problem of lumps, for it suggests that flour deposited in hot water is quickly enveloped by a gelatinized layer that limits the diffusion of water toward the dry central core of the lump. But it is inadequate as an explanation, however, because it fails to explain why, if the water diffuses into the periphery of the lumps, it does not ultimately reach the center.

One-Dimensional Lumps

As we have seen, everything is a question of speed. To measure the rate of gelatinization, let's simplify the matter further by making a one-dimensional lump whose center we can observe. Put some flour in a test tube and then pour water over it (if the water is colored one can follow its diffusion into the flour by observing the movement of a distinct boundary). At room temperature the water infiltrates the upper layer of the flour fairly rapidly at first but then penetrates further very slowly, by less than a millimeter an hour. Heating the water causes the granules to swell, so contact with this gelatinized region limits the advance of the colored boundary. In this case the rate of diffusion rises to several millimeters per hour.

It follows, then, that the center of the lumps remains dry because of the slowness of the water's diffusion through the gelatinized periphery. When the water has diffused and swollen the starch granules, binding them together, they form a layer that retards further diffusion toward the center to such a degree that it remains dry (a phenomenon characterized more precisely by specifying the many molecular interactions that take place between the starch granules and the water). In other words, placing a quantity of flour measuring a centimeter in diameter in hot water causes it to become moistened to a depth of one or two millimeters from the periphery, with the center remaining dry longer than the time needed for most culinary purposes.

How can we get rid of these lumps? No fundamental physics is needed to solve the problem. It suffices to break up the lumps—with a whisk, for example—into particles smaller than the thickness of the starchy layer.

Soaking Gelatin

Does this theory of the formation of flour lumps apply to other types of lumps? Cooks know that leaves of gelatin must soak in cold water before being used in a hot liquid. Failure to observe this rule creates strings that are as difficult to eliminate as lumps of flour in hot water. Are these strings likewise composed of a dry center and a sheath in which the water and gelatin molecules are mixed? To find out, first measure the water's rate of diffusion in cold gelatin by placing a small amount of coffee grounds on its surface. A hemispherical colored zone extends outward from this point at a rate of only about a centimeter a day.

Next, let's repeat the experiment we conducted with the flour, only this time replacing the flour with gelatin. One then observes that the boundary of colored water sinks into the gelatin only very slowly. But a new phenomenon now appears: Under the layer of gelatinous solution, the gelatin that is untouched by the water has melted like butter.

Similarly, in a hot broth, a sheet of gelatin that has not been soaked beforehand has trouble dissolving and conserves a solid central part that melts from the heat. Its molecules stick to one another, forming the dreaded strings. In a sheet that has been soaked long enough to allow the water to gradually penetrate into the center, on the other hand, heating causes the gelatin to dissolve without strings.

41

Foams

The stability of foams depends on the arrangement of the proteins at the interface between the water and air.

FOAMS—LOW IN FAT BECAUSE they are essentially composed of air—first came to prominence with the rise of *Nouvelle Cuisine* in France in the 1960s and then gained broader popularity as a consequence of the growing interest in lighter foods on both sides of the Atlantic. Today, with the advent of molecular gastronomy and, in particular, the fame of the Spanish chef Ferran Adrià, they are very fashionable among gourmets. In the early days foams were obtained by vigorously beating egg whites, but the variety of eggs combined with ignorance of the optimal conditions for making foams led to irregular results, a fatal handicap from the point of view of the food processing industry. Physicochemical analysis of protein foams has yielded a better understanding of the relationship between the composition of proteins and their foaming properties.

Composed of air bubbles separated by liquid films, foams retain their form only if the liquid forming the walls of the bubbles does not subside or if these walls are able to support themselves despite the draining away of the liquid. Beating an egg white reveals one of the conditions of stability, namely, that the air bubbles must be sufficiently small that the surface forces are stronger than the forces of gravity, which cause the water to fall and the air to rise.

By comparing the layers formed by the various proteins along the boundary where water and air meet, Roger Douillard and Jacques Lefebvre at the Institut National de la Recherche Agronomique station in Nantes and Justin Tessié in Toulouse showed that the stability results both from the interactions of the proteins present in the walls of the liquid films that separate the bubbles and from the viscosity of these films.

A key parameter in the study of foams is the interfacial tension of the protein films at the air–water boundary. The physical chemists measured this tension in terms of the force needed to extract a very clean platinum blade immersed in a solution covered with a protein film: The greater the amount of liquid that coats the blade, the greater the force needed to extract it. The proteins modify interfacial tension because they consist of chains of amino acids, with hydrophilic (water-soluble) parts and hydrophobic (insoluble) parts. Arranged at the water–air interface in such a way that their hydrophilic parts are in contact with the water and their hydrophobic parts with the air, they favor an increase in the surface area common to both air and water and facilitate the formation of foams.

A foam is stabilized by increasing the viscosity of its liquid phase (for example, by adding sugar and glycerol) and, above all, by modifying the drainage properties of the absorbent films. In protein foams these films are rigidified by intramolecular and intermolecular bonds, such as disulfide bridges between the cysteine groups of proteins, and weak bonds (van der Waals forces and hydrogen bonds).

The physical chemists from Nantes and Toulouse were particularly interested in the role of proteins in the formation of foams and sought to analyze the scale of interfacial tension as a function of protein concentration. They knew that very soluble proteins, which are adsorbed to only a small degree on the air–water interface, do little to reduce interfacial tension when their concentration rises. But the behavior of almost all other proteins was difficult to analyze because of their molecular complexity: Not only are proteins polymers, which is to say long chains of amino acids capable of being folded into a ball, but they are also molecules carrying electrically charged groups that interact within proteins and with charged groups of other molecules.

The current theoretical description of polymers, developed by Pierre-Gilles de Gennes and Mohamed Daoud at the Commissariat de l'Énergie Atomique in Saclay, predicts that the maximum lowering of interfacial tension ought to be obtained in the case of the molecules that make up the densest adsorbed layers. The presence of electrical charges complicates this prediction, for the greater the charge of a protein the greater its solubility and the less dense the adsorbed layer. By contrast, counterions, which surround the charged protein groups, are supposed to weaken intramolecular and intermolecular attractions and repulsions.

To test these predictions, the physical chemists from Nantes and Toulouse limited their attention to caseins, which, because they do not form a ball, cannot be unfolded (or denatured) during the formation of foams—one less mechanism to take into account. Comparison of various proteins revealed that for globular and nonglobular proteins alike, interfacial tension increases with the concentration of foaming proteins. Two populations of proteins were found to successively appear at the water–air interface, where they constitute distinct molecular layers. The first layer, a weak protein concentration, establishes itself on the surface, with the addition of supplementary proteins triggering the formation of a second layer.

The principal difference between globular and nonglobular proteins involves the structure of the molecules in the layer that is in contact with the air: Nonglobular proteins exhibit a single conformation, whereas globular proteins seem to divide into two subpopulations that differ in respect of the number of amino acids adsorbed at the interface.

What lesson can chefs draw from all of this? Perhaps that they should change their ways. There is no reason, apart from tradition, to make foams from the full protein content of the egg. Why not use specific proteins with superior properties? If they were to combine gelatin and water and beat the mixture for a very long time, for example, they would find that they can obtain quarts of a specific kind of foam from only a few sheets of gelatin. Of course, they would have to make sure that the water they use is flavorful.

42

Hard Sausage

Improving the quality of commercial brands requires identifying the molecules responsible for their aromatic qualities.

HARD SAUSAGES SUCH AS FRENCH *saucisson* are traditionally made by adding sugar, salt, saltpeter, herbs, and spices to a mixture of meats that is then put into a casing and left to dry for several months. This method, which does not rely on any fermentation agents (acidifying and aromatizing microorganisms), can produce both the best and the worst results. Insufficient acidification by bacteria naturally present in the meat carries the risk that pathogenic microorganisms will develop. Moreover, spontaneous flora do not invariably impart a pleasing taste to hard sausage.

In seeking to avoid this Charybdis through the use of controlled fermentation, the food processing industry sometimes falls into the clutches of Scylla: Commercial products often are too soft (the drying process often lasts less than a month), and many of them lack the aromatic quality associated with artisanal sausages. This situation is changing, for commercial manufacturers have an incentive to remedy these defects. In France they have charged researchers from the Institut National de la Recherche Agronomique (INRA) station at Clermont-Ferrand with the task of determining exactly what makes a good *saucisson*.

Jean-Louis Berdagué, Marie-Christine Montel, and Régine Talon set out to extract, and then identify, the aromatic components of hard sausages using high-resolution gaseous phase chromatography (whereby a carrier gas conveys the volatile molecules extracted from a sample into a capillary tube that differentially retains, and therefore separates, the constituent elements of the aroma) in combination with mass spectrometry (which identifies the components separated by means of chromatography). Thanks to these techniques it is now possible to characterize both the properties of the meat used in making a sausage and the curing process itself.

The Clermont-Ferrand team observed first that the aroma of hard sausage derives from about 100 organic compounds produced by the action of enzymes present in the meat and the fermentation agents. Subsequent analysis of sausages prepared from various bacterial combinations showed that the flora responsible for maturation play a crucial role in creating aromas.

In one experiment, six mixtures of acidifying bacteria (*Lactobacillus* and *Pediococcus*) and aromatizing bacteria (*Staphylococcus*) were used to produce thirty different sausages (five samples per bacterial mixture). The volatile compounds in these samples were then analyzed and their aromas evaluated by a dozen trained tasters who had agreed beforehand about the terms to be used to describe aromatic characteristics.

In a further step, a statistical analysis disclosed the relationships between aroma and the presence of the various compounds. More precisely, the oxidation of lipids was found to play a preponderant role in determining aromatic quality: A rancid smell is associated with the presence of aldehydes, alkanes, and alcohols, whereas the smell of good hard sausage is associated with the presence of methyl ketones and methyl aldehydes. Finally, the degradation of sugars favors the development of vinegar odors, created by acetic acid, or of butter aromas, created by 2,3-butanediol.

These studies revealed a situation analogous to the what one finds in the case of yogurt, where the quality of the final product depends on the use of bacterial strains from wholesome milk. Making aromatically rich sausages likewise depends on the quality of the strains used in the maturation process.

Packaging and Aromatization

Further studies carried out by the same team in Clermont-Ferrand in collaboration with Christine Viallon showed that both the length of the curing process and the type of packaging affect the aromatic quality of sausage. Prolonged drying often leads to a loss of aroma because water evaporation carries away with it the volatile compounds. Sausages harden as a result of drying, however, and the progressive concentration of salt brings out their flavor. The INRA team wondered whether wrapping the sausage in plastic film would prevent the loss of aroma through drying. They found that the use of film both limits drying and strongly modifies the degradation of sugars, with the result that products wrapped in this way are less sour and have a more buttery taste.

Finally, the mechanisms of aromatization were also studied. The Clermont-Ferrand researchers were able to detect traces of pepper (terpenes), garlic (sulfur molecules), and brandy (esters formed by the reaction of ethyl alcohol with the fatty acids produced by salting).

Commercial brands of sausage therefore may be expected in the future to reproduce the flavor of artisanal products. Better still, recent research suggests that synthetic aromas can be created that will improve the quality of unsatisfactory bacterial strains. A compound such as 1-octene-3-ol, introduced in the initial stage of fabrication by Didier Roux, a research scientist working with Capsulis S.A., has produced sausages having a delicate aroma of *sous-bois* (wild mushroom with hints of decomposing mossy undergrowth).

43

Spanish Hams

Chemical analysis determines when they are entitled to a protected designation of origin.

IN RECENT YEARS THE PRODUCTION of Spanish hams has risen to almost 30 million units per year. Some of these, obtained from Iberian pigs by traditional methods in the southwest of Spain (more than a million hams a year), have a taste that is so remarkable that the European Union has granted protection to producers in their place of origin. How is this taste achieved? Jesús Ventanas and his colleagues in the veterinary faculty of the Universidad de Extremadura at Cáceres have examined the various stages involved in the long process of preparing these hams and identified the reasons for their distinctive quality.

The traditional method of fabrication has a number of special features that begin with the raising of the pigs. They are of native stock and allowed to roam freely, feeding for the most part on acorns and herbs until their weight reaches 160 kilograms (about 350 pounds). After slaughter in the late fall, when the cool weather prevents meat from spoiling easily, the hams are kept for two days at a temperature of 0°C (32°F). They are then rubbed with salt containing 1% saltpeter and placed in a bed of this salt for a week in a comparably cool environment, 0–4°C (32–39°F). The salt is then wiped off and the hams are left to rest within the same temperature range for two to three months before being dried. The drying lasts for a month and a half in the spring, at temperatures that rise finally to 18°C (64°F). During the summer the hams are stored for another month and a half at room temperature before the final phase of the curing process, which is carried out in cellars for a period of 14–22 months.

Historically, this very long process developed in response to the peculiar character of the climate of the Extremadura region in Spain, but today cold storage rooms equipped with thermostats make it possible to prevent temperature fluctuations. To successfully adapt traditional methods to modern production techniques, producers must be able to determine not only the proper temperature levels but also the appropriate length of time—the shorter the better—for the various stages involved in curing the hams. Recent research has helped them do this.

By the 1970s chemists had shown that products formed by the degradation of proteins (molecules formed by the linkage of amino acids) and lipids contribute to the aroma of certain foods, notably cheeses. Are these products responsible for the aroma of Spanish hams as well, or are they only the precursors of these aromas? In 1990 the Cáceres team set out to analyze the modification of proteins and lipids at each stage of the traditional curing process. They observed that the amino acids released by the decomposition of proteins during salting subsequently became degraded, forming the precursors of aromatic molecules.

Two types of reaction could account for this transformation. Maillard reactions between amino acids and sugars, which are responsible for the flavors of grilled meat, bread crusts, and roasted coffee, are also produced during prolonged storage of food products, causing them to darken in color. Strecker degradations—reactions of amino acids with acids, such as the fatty acids released during the degradation of lipids—produce aldehydes, which are often aromatic.

The Secret Is in the Curing

In Spanish hams the accumulation of Maillard products increases proportionally with the length of maturation, proof that good hams cannot be obtained without long curing. Moreover, aldehydes, which are formed after salting, also play a role in Maillard reactions, yielding products that retard the rate at which fats become rancid.

More recently Spanish chemists have approached the problem from the other direction by first identifying the volatile molecules and then trying to trace their origin. The most abundant ones were found to be are alkanes, molecules that are composed exclusively of carbon and hydrogen atoms. In Spanish hams they are of two types: linear alkanes, which probably come from the decomposition of lipids, and ramified alkanes, a consequence of the distinctive acorn-based diet on which Iberian pigs feed. Chemistry therefore supports the practice of awarding the protected designation of origin only to hams prepared from animals allowed to graze freely in oak groves.

Another aromatically important class of molecules is made up of the linear aldehydes, which are formed by Strecker reactions and by reactions associated with unsaturated fatty acids that turn rancid. Here again chemistry justifies the highest seal of quality, for the meat of Spanish pigs is marbled with fat that contains many such acids.

44

Foie Gras

It melts less and tastes better when it is cooked immediately after the geese are slaughtered.

BOTH ALSACE AND THE SOUTHWESTERN REGION of France claim to have been the first to invent foie gras, but it is not a new delicacy: The Romans are known to have force-fed geese with figs. The manner in which it is prepared is changing, however. Goose livers traditionally were cooked several hours after the animals had been slaughtered, but a revision of food industry regulations in France has led to a centralization of local slaughterhouses and, with it, the practice of immediate evisceration and cooking. What effect has this had on the quality of the livers?

In the early 1990s, a team headed by Dominique Rousselot-Paillet and Gérard Guy at the Institut National de la Recherche Agronomique (INRA) station in Artiguères succeeded in showing that livers obtained by on-the-spot (*à chaud*) evisceration lost less fat than ones that had been allowed to cool down before cooking. Producers were delighted, but they worried that the new method would have adverse consequences from the point of view of taste. To resolve the matter, which concerns one of the glories of French cuisine, Sylvie Rousset-Akrim and her colleagues at the INRA station in Clermont-Ferrand analyzed the sensory characteristics of canned foie gras prepared under both conditions.

Thirty geese were raised and force-fed at Artiguères. After immediate evisceration the large lobe of the livers was cut lengthwise into two parts. One sample was sterilized at once at 105°C (221°F) for 50 minutes, and a second sample, taken from the other half of the lobe, was prepared following the same procedure, only *à froid,* which is to say after several hours of cold storage. Then the two samples were tested by trained tasters.

Earlier studies comparing goose and duck foie gras revealed a great disparity between livers and helped researchers develop the protocol for later tests. The tasters were instructed to grade the intensity of eighteen aspects of the cooked livers on a scale of 0 to 20: appearance (compact, smooth, veined), odor (chicken liver, foie gras), texture (sticky, compact, firm, tender, granular, plump, smooth, crumbly), taste (sour, bitter), and aroma (defined as a detectable part of flavor: foie gras, chicken liver, rancid). Statistical analysis of the results yielded the desired information.

First, the Clermont-Ferrand team showed that human beings are reliable measuring instruments when one knows how to use them. In many cases it was possible to distinguish the two modes of preparation. In particular, the livers that had been cooked at once were both more veined and more tender, plumper and smoother, less granular, less friable, and less bitter, with a foie gras aroma that was more developed than that of livers prepared after a few hours' delay. The sticky livers were the plumpest and the least crumbly. The compact livers were the firmest and smoothest and, again, resisted crumbling. Finally, the most granular ones were the most friable and also the least plump and smooth.

More Tender, More Fragrant

Similarly, the statistical analysis showed a correspondence between odors and aromas, the aromas of chicken liver and rancidity being opposed to the aroma of foie gras. Acidity varied inversely with the intensity of the aroma of foie gras. But to a greater degree than either odor or taste, the various aspects of texture were positively correlated with aroma: In addition to smoothness of appearance and the odor of foie gras, tenderness, compactness, plumpness, and smoothness were associated with the aroma of foie gras.

Without making any prior assumptions, the tasting distinguished two groups. The first consists mainly of livers prepared *à froid,* which have a granular, friable texture and an aroma of chicken liver with a note of rancidity; the second group is composed of livers prepared *à chaud* (along with a few livers from the first group), which are plumper, more tender, and smoother.

What accounts for this division? The fact that the lipid melting rate (the rate at which fatty matter is released during cooking) is far higher for livers prepared *à froid* (21%) than for ones prepared *à chaud* (9%) suggests that rapid removal and cooking of the liver after slaughter prevents the postmortem change of hepatic tissue, limiting membranous lesions and the escape of lipids.

The melting rate depends on the mass of the livers in the case of preparation *à froid* but not in the case of preparation *à chaud*. And no matter which mode of preparation is used, warm or cold, compactness and firmness vary inversely with the weight of the liver: The heaviest ones were found to be less compact and less firm. But when the melting rate increases, as in the case for the livers prepared *à froid*, the foie gras seemed more granular, friable, rancid, and also less smooth, tender, compact, plump, and aromatic.

Producers have no reason to worry, because the new method of cooking livers immediately after the geese have been slaughtered does not harm their product—quite the opposite. Cooks will draw the same conclusion: The best foie gras is made from livers obtained directly from the slaughterhouse.

45

Antioxidant Agents

Aromatic plants prevent the oxidation of dietary fats.

BUTTER AND OTHER FOODS CONTAINING FATS turn rancid on contact with air as a result of a chain reaction: the autoxidation of fatty acids. Arresting this degradation, which produces disagreeable tastes and odors and creates free radicals, has long been an elusive goal. Eliminating oxygen from foods and to protecting them from light is not enough. Antioxidant compounds are also needed to combat the precursors of autoxidation that are already present.

Some foods naturally contain antioxidant compounds that protect them from turning rancid, such as the tocopherols (vitamin E) found in virgin olive oil and the ascorbic acid (vitamin C) of lemon. To extend the shelf life of their products, food processing companies began using these natural compounds more than twenty years ago while seeking ways to synthesize substances with greater antioxidant action. However, consumers feared the unknown toxicity of synthesized products, causing research to be confined to natural compounds.

Understanding the autoxidation of fatty acids nonetheless is important for studies of such compounds. This reaction occurs when light breaks the –C–H bonds of a lipid by forming unstable –C• free radicals that react with the oxygen in the air to form other –COO• free radicals. These radicals then react with other –C–H bonds to create a new –C• radical that propagates oxidation.

Phenols, the antioxidants used by the food processing industry, are molecules containing a benzene ring composed of six carbon atoms at the apices of a hexagon, at least one of which is bound to an –OH hydroxyl group. The antioxidant activity of phenol acids and their esters depends on their structure and, more particularly, on the delocalization of the electrons in their aromatic core: Some of their electrons are shared by all the atoms of the benzene ring. When these compounds react with the free radicals formed by the autoxidation of fats, they are transformed into free radicals, but they remain stable because the nonmatching electrons are delocalized, which limits their reactivity. Thus the propagating reaction is blocked, with the result that foods are prevented from turning rancid.

Despite its well-defined purpose, research on antioxidants was a tedious business because the reaction needing to be avoided appears only slowly; fortunately, it takes time for food to go bad. Systematic investigation using traditional tests of rancidity, which lasted several days, was impracticable because it took much too long.

A new test devised in the 1990s by Hubert Richard and his colleagues at the École Nationale Supérieure des Industries Agricoles et Alimentaires in Massy resolved the problem by indicating the antioxidant power of a compound in a matter of only a few hours. They bubbled oxygen into a lipophilic solvent, dodecane, in which methyl linoleate (the fat to be tested) and the candidate antioxidant were dissolved at 110°C (230°F). Gas chromatography revealed that half of the methyl linoleate is oxidized in three hours. The antioxidant power of a compound is measured in terms of its ability to lengthen this half-life.

The new faster method was then used to elucidate the chemical characteristics of antioxidant compounds (in order to predict which ones would be the most effective) and to determine which aromatic plants are the best sources of these compounds. Comparison of several plant phenol acids with four antioxidants commonly used in the food processing industry—butylated hydroxyanisole (BHA), butylated hydroxytoluene (BHT), 2-tertbutylhydroquinone (TBHQ), and propyl gallate—disclosed the strong antioxidant effect of many compounds. In particular it became apparent that the antioxidant activity of a given molecule seems to depends on how many –OH hydroxyl groups it has and on the degree of stabilization created by the delocalization of electrons. These findings made it easier to predict the antioxidant efficiency of different compounds and to determine in which natural substances the most potent antioxidants are present.

Tests devised at the Massy laboratory for analyzing various extracts from aromatic plants confirmed the long-suspected antioxidant activity of rosemary, sage, cloves, thyme, oregano, ginger, and capsicum, but not of nutmeg. The extracts from rosemary, sage, cloves, and ginger exhibited a degree of activity similar to that of alpha-tocopherol, roughly one-tenth that of BHA and of gamma-tocopherol. Although neither pepper, parsley, celery, Indian celery, nor basil was found to display antioxidant properties, contrary to what earlier studies suggested, benzoin and vanilla act in a way similar to that of alpha-tocopherol. These two plants contain vanillin, which first attracted interest as an antioxidant in 1989.

The most promising plants undoubtedly are rosemary, cloves, sage, ginger, and benzoin. Sage may be a particularly good source, for it has been shown to contain six powerful antioxidant compounds: carnasol, carnosic acid, and isorosmanol, all present in large quantities, as well as rosmadial, rosmanol, and epirosmanol.

46

Trout

Variable quality has prompted physical chemists to take a closer look.

WHAT MAKES SEA TROUT TASTE SO GOOD? Is the pink color of its flesh a sign of quality? How should it be cooked to satisfy gourmets? And how should it be raised to suit processing requirements, particularly in connection with smoking, a technique in which French companies have long specialized?

Although France is among the world leaders in trout production, the study of trout aquaculture is almost thirty years behind the study of beef and other kinds of meat. One of the chief problems is inadequate genetic selection, which causes unwanted variability in the quality of farm-raised fish. Benoît Fauconneau, Michel Laroche, and their colleagues at the Institut National de la Recherche Agronomique stations in Rennes and Nantes, in collaboration with the Institut Français de Recherches sur la Mer and several private companies, have been exploring the physicochemical characteristics and sensory properties of trout filets in an attempt to determine the causes and effects of differences in quality.

It was known that the gustatory qualities of the large salmonids and their susceptibility to treatment by means of techniques such as smoking and marinating depend principally on the concentration of lipids—molecules that dissolve aromatic compounds and that also can communicate tastes when they are heated or oxidized—in the flesh of the fish. Preliminary analysis of fario trout (*Salmo trutta*) raised in the sea by Elsamer S.A., in collaboration with food technologists in Camaret, revealed that the filets obtained from these fish contained 65–70% water, 20–24% proteins, and 2–12% lipids.

The pink color of trout filets, which results from the presence of carotenoid molecules (astaxanthin and canthaxanthin) contributed by food, seems not to be a reliable indicator of quality. Despite some genetic variability in the fixing of these pigments, the color varies mainly according to whether the diet of the fish contains these carotenoids (which cause their flesh to turn from yellow to pink).

Why, then, does the shade of color vary between fish born of the same parents? The answer undoubtedly has to do with the fact that young fish mature at different rates. Indeed, studies have shown that their muscles gain in lipid content in the course of growth: The bigger (and therefore more rapidly developed) the fish, the fattier it is. Because the red component of their characteristic color increases with lipid concentration, it also increases with the rate of growth. This means that a trout that eats a great deal, which is to say one that fattens rapidly, absorbs more pigments than a small fish. In other words, the quality of smoked trout filets would be uniform if the rate of growth could be controlled.

The physical characteristics and sensory qualities of fish depend on the macroscopic organization of their muscle cells, which is very different from that of meats. Muscle cells in animals consist of very long fibers sheathed in collagen, a protective protein, and collected in bundles, which are themselves sheathed in collagen, and so on. The cooking of meat therefore involves a delicate compromise between hardening, which results from the coagulation of the proteins contained in these cells, and tenderizing, a consequence of separation and dissociation of the collagen molecules.

Tender to Cook

In the case of fish, by contrast, the cooking time must be short because their flesh contains little collagen. The muscles are not individually sheathed but are grouped together in sheets of which only the surface is supported by collagen. It is the lipids, localized within the muscle sheets in anatomically distinct adipose tissues, that play a predominant role in holding the sheets together and therefore in determining the texture of the flesh. Tests of resistance to compression have shown that the flesh of trout is firmer than that of other freshwater fish such as carp and catfish.

To understand how trout filets are modified during cooking, the Nantes researchers began by comparing two filets obtained from the same fish: One was characterized at once in its raw state, and the other was analyzed after poaching, at temperatures ranging from 10°C (50°F) to 90°C (194°F), depending on the experiment (the filet was placed in a sealed bag and the temperature increased by 1°C per minute, then rapidly lowered by immersion in ice water).

These studies showed that although cooking did little to change the chemical composition of the filets, it did increase their mechanical resistance as a direct result of the temperatures to which they were heated, which caused the muscle proteins to coagulate. The overall composition of the filets was largely unaffected because their constituent elements were lost in the same proportions in the cooking juices. Because the juices contained a substantial quantity of lipids, the red color contributed by dissolved pigments diminished—all the more so because the luminosity of trout flesh increases as a result of protein coagulation. All told, the loss of matter in the form of liquid, amounting to 10–20% of the total mass, increased as the poaching temperature rose.

Research continues into the effects of cooking (one preliminary result is that trout filets seem not to become softer the longer they are cooked) and the role of diet, which seems to affect the chemical composition of the flesh without much affecting its gustatory qualities. Why should trout not be like fowl, whose taste varies greatly depending on their diet? We do not yet know.

47

Cooking Times

A brief guide to cooking meat so that it will be tender and juicy.

HOW LONG SHOULD A JOINT OF BEEF BE COOKED? The problem is an old one, as we know from Brillat-Savarin's *Physiology of Taste,* and raises the question of what laws govern the various methods that have been devised for cooking foods. Putting aside certain exotic cases, such as the preparation of fish with acids in the Tahitian manner, in which filets are marinated in lime juice, the cook should keep in mind that cooking is fundamentally a transformation of foods by heat.

Heating with Hot Gas

This naturally leads us to ask another question: How is heat most efficiently transmitted for the purposes of cooking? Traditionally foods have been heated by means of gases, liquids, solids, and waves. Let us limit our attention here to the first of these four sets of procedures, which includes smoking, drying, braising, steaming, and oven roasting.

In the case of both drying and smoking the cooking is slow because the temperature of the hot fluid is not much above room temperature. Steaming is more efficient because the food receives both the kinetic energy of the steam and the energy resulting from the condensation of the steam on the food. Nonetheless, the upper limit of temperature in this case is 100°C [212°F]. In an oven, by contrast, the air—filled perhaps with water vapor—is able to reach much higher temperatures, whence the problem of determining the right temperature for cooking, which cooks have resolved empirically by formulas such as "twelve minutes a pound, plus ten minutes for the pot."

What is the basis for such rules? Notice first of all that the maximum thickness of a food determines how much time is needed for it to reach a given temperature throughout. A sausage that is a mile long takes the same time to cook as one that is only a foot long as long as their diameters are equal. On the other hand, a doubling of thickness implies a quadrupling of cooking time because both the distance the heat must travel and the quantity of matter to be heated are doubled. For a spherical body, the cooking time is proportionate to the mass raised to the power of two-thirds, a relationship described by a curve that flattens out after an initially rapid rise. This approximates the old rules, which cannot be applied to foods smaller than a certain size.

Old-fashioned braising, which also relies on a hot gas, is a remarkable operation. Surface microorganisms are destroyed by browning over high heat, and the meat is then placed in a covered baking dish (classically a brazier, nowadays a casserole) and left to cook with "ashes above and ashes below," as the best authors used to say. Although it is less than 100°C (212°F), the temperature of the heating fluid is nonetheless sufficient to evaporate the ethyl alcohol of the brandy customarily used for braising and the various aromatic compounds of the accompanying vegetables. The meat therefore cooks in a fragrant atmosphere, without losing its water, because the ambient temperature remains below the boiling point.

The Fateful Threshold

This technique is fashionable in many restaurants, where it is known as vacuum-packed low-temperature cooking. Foods are sealed in a plastic pouch that has been emptied of air and then poached at a temperature lower than 100°C (212°F). The cooking takes a long time, as in the case of the old braziers, but it makes it possible to prepare dishes in advance. This advantage was so familiar to chefs in earlier times that M. Menon specified no cooking times in his *Science du Maître d'Hôtel Cuisinier* (1750). He knew that as long as foods are cooked over gentle heat, the results do not greatly vary: Meats remain remarkably tender and juicy because their juices have not evaporated.

In earlier times braising was a difficult procedure to master (one had to be careful to guard against the embers suddenly bursting into flame). Today it yields remarkable results as long as one uses a preset oven and keeps in mind a few key temperatures: At 40°C (104°F) meat becomes opaque because the proteins in it, initially folded into a ball, begin to unfold before they coagulate (thus becoming denatured); at 50°C (122°F) the muscle fibers begin to contract; at 55°C (131°F) the fibrillar part of myosin (a protein that, along with actin, is essential for muscle contraction) coagulates, and collagen (a protein that gives meats their toughness) begins to dissolve; at 66°C (151°F) various other proteins coagulate; at 70°C (158°F) myoglobin no longer fixes oxygen, causing the inside of meat to turn pink; at 79°C (174°F) actin coagulates; at 80°C (176°F) the cell walls are ruptured and the meat becomes gray; at 100°C (212°F) water evaporates; and at temperatures higher than 150°C (302°F) so-called Maillard (and other) reactions produce brown and flavorful results.

What is the point of referring to these benchmarks? If the oven's temperature control is calibrated properly, cooks can choose the exact degree of doneness that they want to achieve, without having to depend on unreliable empirical indications and without having to worry about the flare-ups that used to ruin braised dishes in the past. Naturally, meat that is cooked rare has its devotees, but they should not forget that cooking meat at low temperatures favors the proliferation of dangerous microorganisms. Low-temperature cooking is perilous, but the results are wonderful.

48

The Flavor of Roasted Meats

The flavor of roasted meats depends on their fat content.

DO FATS IMPART A DISTINCTIVE FLAVOR to meats? If so, which one? It was long believed that lipids were capable only of dissolving odorant compounds, many of which are water insoluble. They have also been accused of giving meat a bad taste, turning rancid, or oxidizing during cooking. Nonetheless cooks have long known that the flavor of meat is affected by the fats it contains or the fats that are added to it during cooking. Today chemists can confirm that fats play a decisive role in Maillard reactions, whose products are the chief aromatic components of heated foods.

There are hundreds of odorant compounds, which vary according to the type of meat, the age of the animal, its diet, and the mode of cooking. Moreover, compounds present in minute quantities may be aromatically preponderant. One of the principal reactions responsible for generating tastes is the Maillard reaction between sugars (such as glucose) and amino acids. Named after Louis-Camille Maillard, a chemist in Nancy who first identified the reaction in 1912, it contributes to the flavor of bread crust as well as the roasted aroma of meats, beer, and chocolate, among other foods. This reaction also leads to the formation of the dark compounds called melanoidins, which give cooked foods their characteristic color.

Chemists have been investigating the precursors of the volatile compounds of meats for several decades. They first observed that these compounds have a low molecular mass. In addition to the reactive agents typical of Maillard transformations (amino acids and sugars), they found phosphate sugars, nucleotides, peptides, glycopeptides, and organic acids.

The role of lipids, in particular, long resisted explanation. It was known that phospholipids (fatty acids linked to a hydrosoluble group that are very sensitive to oxidation) were responsible for the appearance of fatty and rancid notes, but in 1983 Donald Mottram and his colleagues at the AFRC Meat Research Institute in Bristol, England (now the Institute for Food Research), were the first to observe that they are also indispensable to the development of the characteristic taste of cooked meat. In 1989, their colleague Linda Farmer showed that lipids are involved in the unfolding of Maillard reactions, not only through their degradation products but also on their own account, changing the odorant profiles of roasted meats.

The first studies showed that extracting triglycerides (molecules consisting of a glycerol molecule bound to three fatty acid molecules) from a meat did little to change its odor after cooking, whereas eliminating phospholipids replaced its characteristic aroma with one of roasted meat and biscuit. It is thought that triglycerides are scant in polyunsaturated fatty acids, which gives them a relative degree of chemical stability; many phospholipids are rich in polyunsaturated fatty acids, however, which explains their sensitivity to oxidation. Their hydrosoluble part can react with oxygen as well.

Proof by Reaction

At the Institut National de la Recherche Agronomique station in Nantes, Gilles Gandemer, Anne Leseigneur, and their colleagues studied the role of phospholipids in triggering Maillard reactions in simplified systems. To an aqueous solution of cysteine (an amino acid chosen because it contains a sulfur atom and creates molecules crucial for the formation of the aroma of cooked meat) and ribose (a sugar known for its activity in cooking that can be released in nucleotides) they added either fatty acids found in phospholipids (linoleic acid, palmitic acid, and ethanolamine) or the principal phospholipids in meats (phosphatidylcholine and phosphatidylethanolamine), producing concentrations of various molecules comparable to those found in meats. These mixtures were then heated to 140°C (284°F).

Because the products of Maillard reactions are too numerous to be tested in a controlled way, chemists have sought instead to study changes in chromatographic profiles, focusing on the heterocyclic compounds, which have a meaty taste, and on the products of lipid oxidation. Observing the appearance of new peaks on the chromatograms and the falling off from certain peaks generated by systems modeled without lipids, the chemists were able to confirm that phospholipids have a greater effect than triglycerides. They also showed that the aromas of cooked meat caused by phospholipids arise mainly from two effects: a fatty note created by the presence of carbonylated compounds (which contain the $C=O$ chemical group), the result chiefly of the oxidation of fatty acids, and the interaction of lipids and their degradation products with the direct and intermediate products of Maillard reactions, which leads to the synthesis of a few new molecules and a reduction in the formation of other compounds.

It was also known that the nonvolatile products of Maillard reactions impede the oxidation of lipids. Further analysis showed that the odors of the modeled systems resulted more from a disturbance of Maillard reactions than from lipid oxidation. Although lipids do not come into contact with compounds dissolved in the aqueous phase, phospholipids, because of their polar head, are partially soluble and can react with the intermediate products of Maillard reactions.

49
Tenderizing Meats

Why a meat that is well suited to boiling is not good for roasting.

MEAT IS AGREEABLE TO EAT ONLY WHEN it has been aged for a sufficient period of time. After an animal is slaughtered its meat begins to toughen (for twenty-four hours in the case of beef). This toughness can be reduced by as much as 80% by aging, which lasts for several days (ten in the case of beef). Can this period be shortened, or is it at least possible to determine the minimum amount of time needed to preserve carcass and muscle so that a given cut of meat will be properly tenderized? Ahmed Ouali and his colleagues at the Institut National de la Recherche Agronomique (INRA) station for meat research in Clermont-Ferrand analyzed the characteristics of muscle tissue in order to predict the length of time needed for aging different kinds of meat.

The first stage in the transformation of animal muscles into meat is the onset of cadaveric rigidity. Muscle cells continue to contract and relax immediately after death because they still contain adenosine 5′-triphosphate (ATP), a molecule that stores energy. The chemical cycles of muscle cells regenerate ATP for a certain time, but when it is produced only by the degradation of glycogen (which serves as a reserve supply of glucose) the muscle is no longer able to relax and remains in a contracted state.

During this phase the degradation of glycogen and glucose produces lactic acid, whereas the degradation of ATP releases phosphoric acid, with the result that the muscles are acidified. The swiftness of acidification depends chiefly on the type of muscle: Red (or slow-contracting) muscles, which derive their energy from oxygen carried by the blood, are acidified less and less quickly than white (or fast-contracting) muscles, which do not consume oxygen, making them more vulnerable to alteration by microorganisms.

Tenderizing and Degradation

The conditions under which cadaveric rigidity occurs determine the course of the next phase, tenderizing, which probably results from the degradation of structural elements. A distinction has long been made between the sort of toughness associated with collagen (the protein that sheathes muscle cells, grouping the muscle cells into bundles and the bundles into muscles) and the sort associated with myofibrils (the proteins responsible for the contraction of muscles). It has recently been observed that collagen, which is mostly unaffected by the tenderizing process, provides an index, or baseline, for measuring toughness. Collagen varies according to its concentration in muscles and determines the preferred cooking method for different cuts of meat. Pieces with high concentrations of collagen are best boiled, whereas ones with low concentrations of collagen are better suited to roasting.

Two types of mechanisms seem responsible for tenderizing myofibrils. Certain proteolytic enzymes decompose proteins, breaking down the filaments and fibrils, and an increase in osmotic pressure dissociates the constituent proteins of the filaments.

The INRA biochemists studied three groups of enzymes capable of degrading myofibrillar proteins: cathepsins, calpains, and a complex of proteins known as proteosomes, less well known than the other two because it has only recently been discovered. The activity of these enzymes in muscles depends on acidity, the concentration of calcium ions and ATP, and so on. In living animals it is limited by various inhibitors that prevent the decomposition of muscles, but enzymatic regulation is suppressed after slaughter, largely as a consequence of acidification. Furthermore, the increase in the osmotic pressure of muscle cells that occurs after death facilitates and reinforces the action of the enzymes: The accumulation of small molecules and free salts in the intracellular liquid dissociates the protein complexes, permitting the proteolytic enzymes to penetrate to their substrates more easily.

The sensitivity of myofibrils to proteolytic enzymes greatly varies according to the type of muscle. The myofibrils in red muscle, for example, differ greatly from the ones in white muscle. These differences have to do not only with the identity of the myofibrillar proteins but also with the structure and extension properties of the myofibrils themselves: The more rapidly the muscles contract, the more rapid their enzymatic alteration. This observation explains, at least in part, the well-known relationship between the age of cattle at the time of slaughter and the tenderness of their meat after aging. The aging time ranges from 4–5 days for calves to 8–10 days for steers because the muscles of the older animals are redder than those of the younger ones.

Understanding these mechanisms will allow government researchers and commercial food technologists to concern themselves with the tenderness and, more generally, the quality of meat and to incorporate these properties in the criteria they apply, which today are geared mainly to shortening the time needed for animals to reach maturity. One of the chief objections to this practice is that the very white meat of animals subjected to artificially accelerated growth is also less flavorful and less juicy than the meat of animals from the same breed that have been raised by traditional methods. What is needed above all is patience.

50
Al Dente

The right way to cook pasta.

ANYONE WHO PUTS SPAGHETTI IN HOT WATER for ten minutes or so and expects a good result is bound to be disappointed. Simple though it is, the cooking of pasta raises a number of questions. The first has to do with salt: Must it be added to the cooking water and, if so, why? Is it really necessary to add oil to the cooking water? How can pasta be prevented from sticking?

At home one can quickly make good pasta from scratch by mixing flour (usually made from wheat, but corn or chestnut flour may also be used), a bit of salt, water, oil, and eggs. Long kneading gives body to the pasta, which is then rolled and cut up before being cooked for three to six minutes. During cooking the starch granules absorb water and expand, and the proteins in the egg and flour form an insoluble network that binds the starch granules tightly together, limiting the extent to which they are washed into the cooking water.

Cooks can prevent homemade pasta from sticking by increasing the proportion of egg. If the protein network is formed before the starch swells up, the pasta remains firm during cooking and doesn't stick; if the starch swells up before the protein network forms and the pasta is cooked, part of the starch (chiefly one of two types of molecule called amylose) has time to diffuse in the cooking water, so that the surface of the pasta is coated with the other type of molecule (amylopectin) and its strands stick together. After straining, a chunk of butter or a bit of olive oil will keep the hot pasta from sticking on the plate.

To improve commercial manufacturing techniques, Pierre Feillet, Joël Abecassis, Jean-Claude Autran, and their colleagues at the Institut National de la Recherche Agronomique (INRA) Laboratoire de Technologie des Céréales in Montpellier sought to determine which proteins give pasta its distinctive culinary qualities.

Hard-Grain Wheat Gluten

Commercial producers make pasta from hard-grain wheat. In the absence of egg the protein network is formed by proteins in the wheat and, more precisely, in its gluten. If one takes the mass obtained after kneading flour and water for a long time and then rinses it under a stream of running water, the elastic matter that remains is composed of gluten proteins. Because this substance is more abundant in hard wheats than soft ones, laboratories such as the one in Montpellier are interested in the composition and genetic variability of hard-wheat proteins and in methods of making of pasta that favor the formation of a protein network.

The quality of a good commercial pasta is judged by its yellow-amber color and its culinary properties, which is to say the likelihood that it will not stick after cooking (or even after being slightly overcooked). Plant geneticists therefore have looked to develop hard wheats with firm and elastic gluten. The INRA biochemists showed that this latter characteristic is associated with the presence of a particular protein, gamma-45 gliadin, common to varieties of hard wheat that are rich in glutenins of low molecular mass.

The Montpellier team also investigated the optimal conditions for making pasta and showed that drying it at a high temperature (about 90°C [194°F]) assists the formation of a network of proteins that is more rapidly insolubilized during cooking. This heat must be applied at the end of the drying process in order not to damage the starch granules. Kneading the dough and pulling it through an extrusion press with the aid of an Archimedean screw must also be done in such a way that these granules are preserved. The application of high temperatures acts on the color because it inactivates both lipoxygenases (enzymes that destroy yellow pigments) and peroxidases (enzymes that darken organic material).

Oil, Water, and Acidity

How, then, should pasta be cooked? The first thing to keep in mind is that the proportion of proteins must be high. If hard wheat is not used then one must add eggs to develop the gluten network or else patiently work the dough and carefully roll it out, using enough water to hydrate the proteins so that they are able to bind together. Whatever its composition, pasta must be put into boiling water so that cooking time is reduced and loss of starch content minimized.

Is there any reason to add oil to the cooking water? Batches of spaghetti that have been overcooked, either with or without the addition of oil, show no differences with regard to stickiness as long as the pasta does not pass through the surface layer of oil at the end. Oil is useful mainly because it coats the pasta when it is removed from the water, after cooking. Adding a knob of butter or a squirt of olive oil to the dish at the table produces the same result.

Finally, the cooking water has its own role to play. Jacques Lefebvre at the INRA station in Nantes has shown that the more proteins the water contains, the less amylose the starch loses during cooking. Therefore pasta should be cooked in a rich broth. Moreover, the Montpellier team demonstrated that cooking pasta in mineral water increases the loss of starch content and therefore stickiness as well, whereas pasta cooked in slightly acidified water (through the addition of a tablespoon of vinegar or lemon juice, for example) preserves a satisfactory surface state, even after overcooking. Proteins in water with a pH of 6 have an electrically neutral form, allowing them to combine more easily and form a network that efficiently traps the starch.

51

Forgotten Vegetables

The introduction of novel vegetables requires extensive research.

A HOST OF UNFAMILIAR VEGETABLES—Japanese artichokes, pepinos, Cape gooseberries, Peruvian parsnips, tuberous chervil, sea kale, skirret—are now available to enliven the diet of those who are tired of carrots, leeks, and potatoes. Jean-Yves Péron and his colleagues at the École Nationale d'Ingénieurs des Travaux de l'Horticulture et du Paysage in Angers are studying forgotten or unknown vegetables with a view to improving their reproduction and cultivation. Not only are food lovers eager to try these novel varieties, but economists also welcome the interest of the Angevin agronomists in promoting diversification because saturation of the market for garden produce diminishes the profitability of capital investment. And from the point of view of farmers, cultivating a new or neglected vegetable sometimes is necessary if they are to increase demand for their products.

In Search of Ancient Agronomists

The introduction of new or forgotten vegetables entails a number of related studies: Producers need to be persuaded of the viability of new crops in both horticultural and commercial terms, and consumers have to be convinced of their gastronomic and nutritional interest. It took decades to bring the first Japanese artichokes and Peruvian parsnips to markets in France and a few restaurants. Tuberous chervil and sea kale will not be slow to follow.

Sometimes the reintroduction of a vegetable is made easier by older agronomic studies of the vegetable itself or, when no study has yet been carried out, members of the same family (the various cabbages in the case of sea kale; members of the parsley family, such as celery, in the case of tuberous chervil, and so on). Tuberous chervil, which grows wild in the east of France, was first identified by French agronomist Charles de l'Écluse in 1846. Nonetheless, despite the efforts of many gardeners of the period, its cultivation was never very widespread. When the researchers in Angers began their work, it was limited to a few truck farms in the Dole and Orléans regions.

Tuberous chervil traditionally is planted in the fall, and its roots, similar to carrots, are harvested the next July. Seed-bearing specimens are planted in December. Why is this vegetable largely unknown? No doubt there are a number of reasons: Germination does not take place without prior stratification; the germinating potency of seeds is limited to a year; crop yields are low; the root becomes edible and gastronomically pleasing only when it reaches a certain size, at which point it has a delicate taste similar to that of the chestnut; and the root is very sensitive to the saprophytic mushrooms that develop on surface lesions.

Seeking to remedy these two last defects, the Angevin researchers studied the genetic variability of the species by crossing wild plants with greenhouse specimens and by means of in vitro cultivation, using plant tissues and growth hormones to develop variants. At the same time they looked for ways to improve cultivation methods. In particular, they wondered how the dormancy period of tuberous chervil seeds could be eliminated and the plant population optimized. Because the mechanisms responsible for inhibiting fertilization in related plants (sometimes associated with the presence in the walls of a ripened ovary of inhibitory substances that are easily removed) were already understood, the agronomists were able to test various methods for eliminating the dormancy period under controlled conditions: variations in temperature, light cycles, application of growth regulators (gibberellin and cytokinin), conservation under oxygen-free conditions, and so on. Finally, they showed that dormancy is embryonic in nature; that its onset occurs during the last stages of the seeds' maturation, once they have lost most of their water; and that it can be blocked only if the seeds have been exposed to a cool, moist environment for eight to ten weeks.

This research has also led to the creation of new plant lines and the selection of a number of commercially promising hybrids. By themselves, however, such studies do not guarantee that the new vegetable will be brought to market. Growing techniques must be explained to farmers and seed manufacturers, and promotional efforts aimed at bringing it to the attention of the public must continue until it reaches buyers' dinner tables. Nutritional studies therefore are needed to determine how a new vegetable is best prepared. For example, tuberous chervil loses starch, gains in various reductive sugars, and develops its distinctive flavors during storage. Recipes created by renowned chefs on the basis of these studies can make a great difference in determining whether the new (or newly reintroduced) vegetable is accepted.

52

Preserving Mushrooms

Modifying the atmospheric pressure under which button mushrooms are packaged helps retain their freshness for a long time.

MUSHROOMS ARE FRAGILE AND HARD TO KEEP. Consumers will put up with wild mushrooms that are a bit bruised, but they want ordinary button mushrooms to be nice and white, with a short stem, small cap, and gills that are covered by a continuous veil, for they know that mushrooms can rapidly change character. A few days is all it takes for mushrooms to darken, for their stems to lengthen, for their ink-black gills to be exposed, and, worse still, for their taste and texture to be denatured.

How can mushroom producers keep their products fresh for longer periods of time? The success of various kinds of ready-to-eat salad greens, washed and packaged under controlled atmospheric pressure, has encouraged food processing firms to take an interest in the problems associated with bringing mushrooms to market. Having shown that the shelf life of baskets of mushrooms could be extended by altering the preservation atmosphere, a team of researchers from the Institut National de la Recherche Agronomique (INRA) station at Montfavet working with food technologists from the Association de Développement et de la Recherche dans les Industries Agro-Alimentaires et d'Emballage-Conditionnement studied various plastic films to determine which one does the best job of sealing in freshness.

Living, Breathing Mushrooms

The researchers began by analyzing button mushrooms kept at room temperature and were able to confirm that cold temperatures reduce microbial and physiological degradation. The usual form of commercial display calls for mushrooms to be stored for a day or two at 2°C (36°F) and then put out for sale at room temperature for a day. This means they will be fresh for only a brief period after purchase: At 11°C (52°F), in 90% relative humidity, mushrooms remain presentable for three to five days, but at 13°C (55°F) this period is reduced to three days.

The development of a system of refrigerated storage and transportation links ("cold chains") for delivering fresh prewashed salad greens to the consumer has made it possible to offer button mushrooms in plastic wrapping that significantly extends their sell-by date. But in order to know what kind of atmosphere is optimal for such packaging, it is necessary to understand how mushrooms change under different conditions.

In the complete absence of oxygen, mushrooms are colonized by potentially dangerous microorganisms such as *Clostridium botulinum*. In the presence of oxygen, on the other hand, mushrooms continue to breathe and change. Since 1975 it has been known that the color and texture of mushrooms depend on the atmosphere is which they develop. In particular, increasing carbon dioxide concentration while reducing oxygen concentration lowers the respiration of fungal cells and, for this reason, retards degradation.

To determine under which atmosphere mushrooms are best preserved, G. Lopez Briones and his colleagues at the Montfavet INRA station measured the effect of various concentrations of oxygen and carbon dioxide on mushrooms stored at a temperature of 10°C (50°F). They observed a correspondence between carbon dioxide concentration and the phytotoxic effect of this gas (detected by the increased respiratory intensity of the mushrooms on being put back into contact with air), which damages the cell membranes and thus favors their exposure to darkening enzymes and their substrates. From the point of view of color, the best atmospheres are those in which the carbon dioxide and oxygen concentrations are lower than 10%.

By contrast, texture is best preserved when the carbon dioxide concentration is higher than 10%, for mushrooms have a superstructure that resists modification. In specimens preserved for a week at 10°C (50°F), in an atmosphere containing 15% carbon dioxide, the veils are not ruptured, preserving the button form preferred by French consumers; the higher the carbon dioxide content, the longer the cap remains closed. Finally, excessive humidity causes rapid degradation, so if you put button mushrooms in the refrigerator, don't close the bag.

Slowing Maturation

A compromise must be found: The carbon dioxide concentration should be not too high, or the mushrooms will not remain white, nor should it be too low, or they will develop too rapidly. Carbon dioxide concentrations between 2.5% and 5% and oxygen concentrations between 5% and 10% appear to be optimal.

How can such atmospheres be created? The Montfavet team compared new microperforated polypropylene films and stretchable polyvinyl chloride films. Mushrooms were placed in baskets, some wrapped and some not, and stored at temperatures between 4°C (39°F) (the legally mandated temperature for cold storage of ready-to-eat salad greens) and 10°C (50°F) for eight days. After eight days of storage at the high end of this range the veils on 85% of the unwrapped mushrooms had opened, whereas maturation was slowed at all temperatures in those that were covered with both types of film. But the old polyvinyl chloride type, being less permeable to humidity, was more successful at retarding development. The challenge now facing researchers is to maximize this impermeability.

53
Truffles

European black truffles are all of the same species, but genetic analysis shows that Chinese truffles are something quite different.

THE BLACK DIAMOND! An immense amount of ink has been spilled in singing its praises. No food writer fails to mention its appearance on a menu, and no chef neglects to feature it when he aims for stars. For centuries the merits of the various black truffles that grow in Western Europe have been debated. The black truffle of Périgord is recognized have a quite different taste from the one found in Burgundy, and naturally the truffles found in France are claimed by the French to be far superior to those of Spain and Italy. Can science provide an objective basis for these opinions?

In Europe there are ten sorts of truffles, which is to say mushrooms of the *Tuber* genus. The black truffle, also called a Périgord truffle (*Tuber melanosporum*), is harvested principally in Spain, France, and Italy, but its gastronomic qualities vary from region to region. Michel Raymond and his colleagues at the Centre National de la Recherche Scientifique, the Institut National de la Recherche Agronomique (INRA), and the Institut pour la Recherche et le Développement in Montpellier sought to determine whether these differences have a genetic basis.

More than 200 samples from various regions in France and Italy were analyzed. The Montpellier biologists studied satellite DNA sequences, which differ substantially between species of the same genus, and observed no genetic variability in the samples of black truffles. They also compared black truffles with summer truffles (*Tuber aestivum*) and Burgundy truffles (*Tuber unciatum*).

Differences were found between the black and summer truffles, as expected, but the species boundaries between summer and Burgundy truffles turned out to be fuzzy. Subsequent studies confirmed that the latter two types must be considered distinct varieties—what biologists call a species complex.

How is the genetic homogeneity of the *Tuber melanosporum* species to be explained? It is believed that the last Ice Age trapped a small population of black truffles next to the Mediterranean, along with the trees on which they develop (the mushroom forms a subterranean root network, of which the truffle is only the reproductive organ). Because the black truffle matures during the winter, from November to February, its propagation was confined to the most southerly zones. During the later climatic warming, the black truffle is thought to have recolonized the regions where its favored trees first developed, as weather conditions permitted. Ten thousand years would have been enough time for the species to reestablish itself in southern Europe but not enough time for it to evolve.

During the same period, by contrast, summer and Burgundy truffles, which mature in spring and fall, respectively, are thought to have been confined to a more northerly area. Certainly the fact that they are found in countries further to the north and east of France proves that they are able to tolerate colder climates. The current genetic diversity of species therefore results from the fact that present-day truffles are descended from a numerous and varied population.

A Himalayan Truffle

Genetic studies completed these results by illuminating the problem of Chinese truffles, which every year swindlers try to pass off as French truffles. In 1996, Marie-Claude Janex-Favre and her colleagues at the University of Paris–VI studied the Chinese truffle, which initially had been assigned to the species *Tuber himalayense, Tuber indicum,* and *Tuber sinense*. These truffles come from the foothills of the Himalayas, where they are harvested at an altitude of about 2,000 meters (6,500 feet), at least a dozen centimeters (four or five inches) below ground. They are easily confused with the French truffle, and the cost of harvesting them is far lower.

The Chinese truffles have a very irregular, bumpy surface. Reaching as much as 7 centimeters (almost 3 inches) in diameter, they are covered with low scales displaying an inverted pyramid form with a square base. This general appearance is almost identical to that of a particular kind of Périgord truffle that is covered with large flat scales and devoid of marked protrusions. The spores of the two types of truffles nonetheless look different under the microscope. Does this difference in appearance result from a difference in developmental conditions or from speciation? Genetic studies carried out by Delphine Graneboeuf and her colleagues at the INRA station in Clermont-Ferrand have established that they belong to different species. The mild taste of Chinese truffles therefore is a consequence of genetic rather than environmental factors.

In the case of black truffles, although environmental factors are now known to be responsible for differences in gastronomic quality, science has not yet explained how the soil and climate exert their influence. Biologists are busy trying to find out.

54
More Flavor

Odorant molecules are trapped so that they may be better perceived.

SOME CLASSIC RECIPES APPEAR PARADOXICAL. To make a *salmis de canard,* for example, one removes the breasts and thighs of the duck and roasts them. The cook then makes a sauce by cooking the scraps and bones of the duck in water together with vegetables and aromatic herbs. Isn't the second step redundant? No, because in the final dish the odorant and taste molecules contributed by the duck are retained for different lengths of time by the meat and the sauce before being released. As a result, the flavor of the dish lingers in the mouth longer.

The problem of trapping odors has received additional attention with the widespread use of extrusion cooking technology by food processing companies. This technology, borrowed from the polymer industry, involves a screw that turns inside a cylindrical barrel, as in a meat grinder, only here the screw is tapered (so the distance between the edges of its blades decreases from the top of the screw shaft to the bottom). A mixture of solids and liquids placed in this apparatus is forcefully compressed—so much so that the water that is suddenly released evaporates, causing the food to be "puffed." Cocktail crackers typically are made using this method.

The puff is achieved at the expense of flavor, however, for the food's odorant molecules are carried away with the evaporated water. Commercial producers often spray aromas on the extruded crackers afterwards. But because this is a costly expedient, they are interested in finding ways to trigger chemical reactions during the extrusion process that will produce odorant molecules or otherwise to trap the volatile molecules.

How can molecules be trapped? In the sauce that accompanies the roast duck, for example, the long cooking of the scraps and bones in water with vegetables has the effect of extracting the gelatin present in the skin, tendons, and bones, producing a decoction that, as chemists very well know, extracts the odorant and taste molecules present in animal and vegetable matter. Odorant molecules therefore are found in two distinct physicochemical environments: in the meat, where they are dissolved for the most part in fats, which are themselves dispersed between the muscle cells; and in the sauce, where they are in a liquid solution. In the mouth these molecules are released in different ways, so that the flavor of the duck lasts longer.

Retention in Solution

Let's try to figure out why this is so. If the duration of a dish's flavor in the mouth has to do with the release of odorant and taste molecules, how can this release be controlled? By controlling the environment of the odorant molecules.

In their pure state these molecules are highly volatile, and the pressure of saturated vapor increases with temperature. The cook is able to vary the degree of volatility by redistributing the odorant molecules into environments of different temperatures. The molecules can also be placed in solution. In this case their volatility depends on the solvent used (whether water, alcohol, or oil) because the molecules bind to a greater or lesser degree with the molecules of the solvent (thus saturated and unsaturated oils differentially retain odorant molecules in solution).

Playing with Molecular Interactions

Cooks can slow down the evaporation of volatile molecules further by putting them in the presence of larger molecules, with which they bond. For example, we know that iodine turns food containing starch blue because the amylose molecules in the starch (long chains composed of linked glucose molecules) wrap themselves around the iodine in a helix. More generally, in water solution, the amylose wraps around hydrophobic molecules, many of which are odorant molecules. Amylose is by no means the only molecule that wraps around in this fashion: Gelatin does the same thing, hence its usefulness in sauces. More generally, well-chosen flexible polymers bond with odorant molecules to retard their release.

Compartmentalization is another, more radical means of retaining molecules. For example, fines herbes (a mixture of fresh chopped herbs used in cooking that typically includes parsley, chervil, tarragon, and chives) release their odorant molecules only when their cells are ruptured by chewing. Emulsions, foams, gels, and pasta are systems of the same type. Similarly, cooks may soon be able to use liposomes, which are sorts of artificial cells created by the assembly of molecules analogous to those of cell membranes. This question is being studied as part of a European Union project devoted to the innovative transfer of technology for culinary purposes.

In addition to microcompartmentalization at the cellular level, there are various forms of macrocompartmentalization. Consider the farces, or stuffings, used in classic cooking, in which aromatic meat and vegetable mixtures are placed beneath the skin or inside the bodily cavities of fowl and other animals. Odors can also be retained if they are made to penetrate foods: The flavor of a meat that is marinated or of a meat that cooks in a fragrant broth is enriched by the odorant molecules of the marinade or broth. Related techniques include decoction, infusion, and maceration.

There are many methods, then, for retaining aromas to one degree or another so that they are released at different moments as one eats a dish. The greatest cooks are able to create waves of flavor that call to mind the peacock-tail effect of the great wines of Bordeaux.

55
French Fries

A new kind of potato for frying, packaged raw, absorbs less oil than frozen fries.

MANY OF THE FRIES SERVED TODAY in restaurants in France come out of a vacuum-sealed bag. Home cooks still prefer to use fresh potatoes because they soak up less oil. Will traditional practice give way to the new technique of packaging sliced potatoes raw under a controlled atmosphere, developed in 1997 by Patrick Varoquaux and his colleagues at the Institut National de la Recherche Agronomique (INRA) station in Montfavet?

In cafeterias and restaurants at least there is no avoiding the superior convenience of precut and processed potatoes because the large volumes consumed take extensive advance preparation. Another complication is that potatoes darken upon exposure to the air once they have been cut up because slicing releases enzymes and associated substrates that otherwise are shut up in separate compartments in the cells of the potato. In the presence of oxygen these molecules react and form brown compounds similar to the ones that cause our skin to tan in the summer.

Before Varoquaux's work, this enzymatic browning prompted the makers of prepared fries to offer precooked products: The potato sticks were peeled and sliced, then dried and deep-fried in oil (often palm oil, cheaper than other kinds), and finally frozen. For the final cooking they could either be deep-fried again, in which case the microscopic fissures created during freezing caused them to absorb a lot of oil; or reheated in the oven, in which case they ended up being too dry.

Atmosphere of Fries

In their search for ways to remedy these defects, the INRA chemists experimented with the idea of packaging raw sliced potatoes under controlled atmospheres—what are now known in Europe as *quatrième gamme* fries—taking care to monitor signs of browning during fabrication.

To minimize browning, a number of steps are followed. First, the potatoes must be carefully peeled, preferably under a stream of water, so that the cellular structure is not damaged. For this reason the blades of the stainless steel knives used to cut the potatoes into sticks must be kept as sharp as possible.

Next the individual sticks are kept at a temperature of about 4°C (39°F) so that the metabolism of the intact cells is slowed down as much as possible. After draining by either centrifugation or ventilation, the potatoes are treated with an inert gas, in the absence of oxygen, in a perfectly sealed packet. In this way the sticks can be preserved for 10 days, still at 4°C (39°F), without alteration (in the course of storage, however, the tissues of the potato accumulate sugars that cause the fries to darken during cooking, by reactions analogous to those that brown the crust of bread). The flavor and texture of fries that are cooked later nonetheless resemble the flavor and texture of fresh French fries: The proportion of oil absorbed is similar, much less than in the case of frozen potatoes that are deep-fried.

Deep-Frying Considered

How should French fries be cooked? On this point cooks are apt to disagree, for each chef has his or her own method. One needs to ask what one is looking for in a plate of French fries and then rationally to examine which procedures allow this expectation to be satisfied.

Few connoisseurs will quarrel with the opinion that good French fries must be tender at the center, with minimal greasiness, and that they should be crispy without being overly brown. To achieve this result we must recognize that deep-frying involves a diffusion of heat from the outside inward, with two principal consequences: the formation of the crust and the cooking of the interior.

Potatoes are composed of cells that contain mostly water and starch granules. When the heat reaches the center of the fries by conduction, some cells are dissociated as the starch granules release their long molecules into the heated cellular water. With the complete evaporation of this water a crust is produced on the surface of the fries.

If one cooks a potato stick into which a thermocouple has been inserted (a more rapid and more reliable way of measuring temperature than with a thermometer), one finds that the interior heats up very slowly: Even when the temperature of the oil is 180°C (356°F), the temperature in the center reaches 85°C (185°F) only after several minutes, for the potato is thermically inert. In other words, if the oil is too hot in the first round of frying, the surface will burn before the inside is cooked.

Conversely, the oil must not be too cold to begin with, for then the crust will be slow to form and the fries will soak up oil. In practice, seven minutes of cooking at a temperature of 180°C (356°F) yields good results for fries measuring 12 millimeters (about half an inch) thick. A second round of cooking in oil heated to a slightly higher temperature, 200°C (392°F), produces perfect fries; remove them from the oil when they have turned just the right golden brown color.

56

Mashed Potatoes

Proteins change the behavior of starch in water.

WHY DO MASHED POTATOES MADE WITH MILK stick less than ones made with water? By showing that proteins modify the thickening and gelatinization of starch, Jacques Lefebvre and Jean-Louis Doublier at the Institut National de la Recherche Agronomique station in Nantes have at last provided an explanation of this venerable piece of culinary knowledge and useful indications for the use of flour in sauces.

Flour and potatoes have in common the fact that they contain a great deal of starch, in the form of granules that contain two sorts of molecules: amylose, which is composed of a linear chain of hundreds of glucose groups, and amylopectin, a polymer that is similar to amylose but has instead a ramified, or branched, structure. In each granule of starch these two types of molecule exhibit a crystalline form.

Whereas the starch granules in flour are exposed to the surrounding atmosphere by the milling of grains of wheat, the starch in potatoes inhabits a watery environment enclosed by their cell membranes. Starch does not dissolve, for amylopectin is highly insoluble in water, and amylose is soluble only in fairly hot water, at temperatures higher than 55°C (131°F).

When one cooks potatoes by putting them in a hot fluid (air, water, or oil) the heat is drawn to the center by conduction, triggering the same sort of expansion as the one that occurs when, in preparing an espagnole, a white sauce, or a béchamel, one pours a boiling liquid (water, milk, or broth) into flour: The water molecules dissolve the amylose molecules and alter the structure of the starch granules, disorganizing the amylopectin crystals and causing the granules to swell. During these transformations the starch granules soften, and dissolving the bulky amylose molecules makes the solution flow less easily. Thus a sauce thickens when flour is added to it, both because the swollen granules get in the way of one another and because the solution in which these granules are dispersed becomes viscous.

If cooks do not always understand the molecular details of the thickening of sauces, they nonetheless know that sauces must be kept hot. For example, white sauces form a gelatinous mass when they cool because a gel forms when the neighboring amylose molecules in the water solution combine with one another, forming a network that traps the water, swollen starch granules, and the various dissolved compounds.

Doublier and Lefebvre showed that the phenomena of swelling up and gelatinization are modified when proteins such as casein are present outside the swollen granules. Caseins, found in milk, aggregate into structures called micelles, which are dispersed in the cellular water and coat droplets of fatty matter. As emulsifying agents they are widely used in the food processing industry to make ice creams, dairy products, custards, and so on.

These proteins reduce the quantities of amylose that leak out of the starch granules and also limit the extent to which they swell up. Caseins subsequently bring about a separation in the water phase: Protein-enriched water droplets separate from the rest of the sauce, which is then enriched by amylose in a continuous phase. This increase in amylose concentration favors its gelatinization.

The Proteins of Mashed Potatoes

How do these mechanisms operate in the making of mashed potatoes? When one cooks potatoes the starch granules are not fully expanded because there is not enough intracellular water to be absorbed by the granules. Mashing the potatoes with a bit of reserved cooking water hastens the incorporation of amylose into solution and the swelling up of the starch granules, with the result that the whole thing forms a sticky mass. On the contrary, mashing them with milk, which contains caseins, limits the swelling of the starch and therefore yields a smoother, more pleasing consistency. This phenomenon also must be taken into account when one thickens sauces with flour, for the gelatin of the sauce stock has the same effect on the flour as milk proteins.

In studying variations in viscosity as a function of the rate of flow, the Nantes researchers provided another hint that cooks ought to find useful. Although perfectly expanded starch granules are thixotropic, which is to say that they are deformed when subjected to flow and therefore form a less viscous solution in the mouth than at rest in the saucepan, sauces thickened with flour, in which proteins limit the degree of expansion, preserve an almost constant level of viscosity: The few amylose molecules that leak out of the starch granules into solution are aligned with the direction of flow, while the unswollen granules are deformed to a correspondingly minor degree.

57
Algal Fibers

Algae contain fibers whose nutritional value is comparable to that of vegetable fibers.

IN PARTS OF THE FAR EAST algae have been used as vegetables in soups and salads since ancient times. In France they serve mainly as a source of iodine and fertilizer and as gelatinizing or texturing additives. Although eleven species of algae were recently accepted as vegetables by the French health authorities, their chemical composition and metabolism are poorly understood. Analysis of the fibers they contain nonetheless has illuminated the sources of their nutritional value.

The modern vogue for fibers began in the early 1970s when the British physician Denis Burkitt discovered a correlation between certain digestive, metabolic, and cardiovascular diseases and low levels of consumption of foods rich in fibers. Fibers are macromolecules that resist digestion by human enzymes. For the most part they make up the cell walls of plants—cellulose, for example, and various other molecules composed of chains of monosaccharides, or elementary sugars. Their chemical complexity dampened the enthusiasm initially aroused by Burkitt's finding, but researchers have now developed sufficiently powerful analytical tools to carry on his work.

Fibers and Digestion

At the Institut National de la Recherche Agronomique station in Nantes, Marc Lahaye and his colleagues have used such tools to study the dietary fibers of marine macroalgae. It used to be customary to classify fibers according to their degree of solubility in response to various enzymatic treatments. Water-soluble fibers—certain pectins (fruit polysaccharides that cause jellies to gel), algal polysaccharides, and certain kinds of hemicellulose—were distinguished from insoluble fibers such as cellulose, other kinds of hemicellulose, and lignin. The soluble fibers, many of which have interesting rheological properties, were thought to reduce the blood concentration of cholesterol and to act on the metabolism of glucids and lipids. Insoluble fibers, on the other hand, seemed to accelerate bowel movements.

Researchers have confirmed the main features of these properties in recent years, and the Nantes team refined the classification of the dietary fibers in algae by modifying a technique of molecular separation known as the gravimetric method, which was used to precipitate macromolecules in various environments, such as water and alcohol, after the elimination of starch and proteins by enzymatic treatment. In 1991 Lahaye and his coworkers used it to determine the quantity of polysaccharides solubilized in environments that approximately reproduce those of the digestive tract.

They went on to apply this method to algae and found that the total concentration of dietary fibers in wakame, for example, can be as great as 75%, as opposed to only 60% in Brussels sprouts, the root vegetable that is richest in fiber. To discover how such fibers behave in the digestive tract, the Nantes chemists first studied *Laminaria digitata,* several thousand tons of which are processed every year in order to produce the gelling agents known as alginates (in 1992 alone some 65,000 tons were recovered and treated by the French seaweed fertilizer industry). Its polysaccharides—glucose polymers known as laminarins—are essentially soluble in very acid environments, whereas alginates dissolve in a neutral environment. The insoluble fibers of *Laminaria* species are principally constituted by cellulose.

On the other hand, the fibers of dulse (*Palmaria palmata*), a common red seaweed consumed by Europeans since the eighteenth century, seem to be continuously solubilized in the successive sections of the digestive system. In the case of sea lettuce and sea hair—two of the eleven species of algae that can now be sold as vegetables or condiments in France—the soluble fibers are xylorhamnoglycuronane sulfates (made from sugars and composed, in particular, of xylose and rhamnose), and the insoluble fibers are principally glucans (glucose polymers). Their solubilization more nearly resembles that of dulse.

Algal Thickening Agents

Improved understanding of the nature and assimilation of these foods has stimulated interest in developing markets for underused algal fibers and for the large quantities of residues that are now extracted from alginates every year. The extraction procedure begins with repeated washings in an acidified water solution, which eliminates the laminarins and fucans and transforms the alginate into alginic acid. Dietary fibers are solubilized in these baths. Then the alga is tossed in a warm basic environment (often with sodium carbonate), and the insoluble part is separated out with the aid of a flocculating agent and air currents. The resulting flocs, or tuftlike masses, constitute a second coproduct consisting mainly of cellulose.

Fibers from vegetable sources—beets, cereals, and fruits—are now incorporated in breakfast products or used as an ingredient in various prepared foods. Algal fibers can serve the same purposes. Looking to the future, research in the chemistry of algal fibers holds out the prospect that processing methods can be developed similar to those used in the milling of wheat and other grains, which depend in large part on an understanding of their elaborate structure. But whereas the chemistry of starch is already well developed, the listing of the elementary sugars that make up algal macromolecules has only recently been completed, and the exact order of their sequence is in many places still uncertain, as is the type of chemical bonds linking the sugars. Now that chemists have decoded the alphabet of algal polysaccharides, it remains for them to learn how to form words and put them together.

58

Cheeses

Commercial protection requires several kinds of analysis.

TO PROTECT THEIR CAMEMBERTS and other raw milk cheeses against legislation that would prohibit the use of raw milk in fabrication, countries such as France must demonstrate the gastronomic superiority of these cheeses by comparison with ones made from pasteurized milk. To do this they must perform a detailed comparison of the chemical composition, texture, and aroma of the different types, which will take quite a while. In the meantime, the Institut National de la Recherche Agronomique and the Comité Interprofessionnel du Gruyère de Comté, in collaboration with food research institutes elsewhere in Europe, have been analyzing cheese with the aid of a very efficient system: human beings.

The chemical analysis of cheeses, as by chromatography, often is insufficient. As a device for detecting trace amounts of molecular compounds, which sometimes are aromatically preponderant, the human nose has no rival. Nor are measurements of mechanical properties very helpful in characterizing certain cheeses that are pasty, crumbly, and heterogeneous.

On the other hand, tasting—an essentially subjective activity—integrates various pieces of sensory information that human beings sometimes have a hard time dissociating. A group of experts in the sensory analysis of cheeses, assembled under the auspices of a European Union program called Food Link Agro-Industrial Research (FLAIR), has sought to set standards for the training of tasting juries and to define norms for characterizing hard and semihard cheeses. Six cheeses awarded a protected designation of origin were tested: Comté (from the French Jura), Beaufort (from the French Savoie), Parmagiano-Reggiano and Fontina (Italy), Mahon (Spain), and Appenzeller (Switzerland).

The Texture of Cheeses

The texture of cheese is a crucial property, as anyone who has eaten a chalky Camembert or a rubbery Gruyère knows. Unlike the organoleptic qualities of shortbread biscuits, gum arabic candies, or puffed cocktail crackers, those of cheeses cannot be described in terms of a single textural characteristic, for the flavor of a cheese depends in complex ways on its overall texture.

The European laboratories participating in the FLAIR study evaluated the texture of the six selected cheeses by comparing their superficial, mechanical, and geometric characteristics in addition to various sensations they produce in the mouth. To specify perceptions and intensities, the researchers characterized the samples in terms of a reference class of basic textures associated with a particular kind of apple, a cracker, a banana peel, and so on.

In the 1960s American food researchers proposed a general classification of the textural properties of foods, but the study of cheeses has since shown it to be inadequate. Which sensory characteristics must be considered in order to define the texture of a food? And how can they be systematically recognized? Specialists in sensory analysis have devised a range of methods for evaluating products: analysis by untrained judges, by trained judges on the basis of predetermined categories, and by trained judges using intuitive criteria. The attempt to establish international norms required a strict methodology. The protocol finally chosen involved predetermined categories with trained judges conducting blindfold tests.

Sensory evaluation consisted of the following steps: looking at the sample, touching it, chewing it, deforming it, and shaping it into a ball before swallowing. Because the overall assessment of texture was to be made with reference to surface, mechanical, geometric, and other objective measurements, each phase of tasting was classified under one of these four categories. The intensity of each sensation was evaluated on a scale of 1 to 7, with a minimum of three basic textures serving as points of reference. Tasting was done at a temperature of 16°C (61°F).

Parallelepipedal slices of cheese having an area of 150 square centimeters (about 24 square inches) and a thickness of 2 centimeters (about three-quarters of an inch) were visually inspected to assess the smoothness of their surfaces and the presence of any openings, tears, crystalline deposits, and droplets of water or oil. Tactile information regarding superficial quality was collected next: Samples were put on a plate, and the judges ran their index fingers over each slice to feel the grain and the degree of moistness. Mechanical characteristics were noted by examining strips 1.5 cm (about half an inch) wide and 5–8 cm (about 2–3 inches) long that had been cut along the grain of the cheese. The judges then evaluated elasticity (by pressing down with the thumb), firmness (by gently pressing down with the teeth), deformability (the maximum deformation achieved before breaking), friability, and adhesivity. Finally, geometric characteristics related to the shape, size, and nature of the particles perceived in the course of chewing (sandiness, granularity, fibrousness, and so on) were analyzed, along with other textural characteristics resulting from complex and residual perceptions such as solubility, impression of humidity (dry or moist), and astringency.

Cultural References

The products used as bases for comparison in evaluating all these characteristics were chosen because they are readily available in European countries and because, in the case of prepared dishes, they are simple to make using standardized procedures. For example, roughness was determined by comparison with the outer surfaces of a Granny Smith apple, a banana peel, a ladyfinger, and a Breton cake; solubility was determined by comparison with a long madeleine, a cooked egg yolk, and a small meringue; and so on.

In the case of hard and semihard cheeses, further study will be needed to examine correlations between sensory descriptions of texture and mechanical and biochemical analyses. New European Union research programs have been proposed for soft cheeses, for there remains a great deal to learn about the reasons for the distinctive character of raw milk products. However, Camembert is not included in these programs, for lack of a commercial sponsor (in Europe the costs of such research fall on industry). For the moment, then, Camembert's superiority is a fact only in the minds of connoisseurs.

59
From Grass to Cheese

Diet contributes to the quality of cows' milk cheeses.

LE TERROIR—THE LAND. For several years food producers have been talking of nothing else, often out of a desire to protect or expand their markets. If they are to be believed, there is a special relationship between a region and its products; no version of a product made elsewhere is as good as the original, which for this reason is the only one to merit a protected designation of origin.

Agronomic analysis has demonstrated the effects of climate, soil, and exposure in the case of wine, but cheese proved to be a trickier proposition. Working with producers in the northern French Alps and the Massif Central, Jean-Baptiste Coulon and his colleagues at the Institut National de la Recherche Agronomique station at Clermont-Ferrand recently succeeded not only in illuminating the relationship between the land and the quality of its cheeses but also in proposing objective criteria for according them certain legal protections.

Taste and Region

What makes a cheese good? The distinctive characteristics of a cheese may result from the physical features of the region in which it is produced, the types of animals that provide the milk from which it is made, and the people who make it. Producers who talk about *terroir* emphasize the importance of the cows' environment and, in particular, of their diet: green grass, hay, and silage (wet grass stored in silos, where it ferments). Does this diet really determine the quality of a cheese?

The first attempts to answer this question, beginning in 1990, involved twenty makers of Gruyère in the Franche-Comté region of France and revealed a significant correlation between the taste of their products and the geographic site of their operations. In other words, one can give a rigorous definition of what "raw milk" means in Franche-Comté. Subsequent analysis of the volatile components of Gruyère made in Switzerland showed that mountain pasture cheeses are distinct from those of the plains: Fragrant molecules belonging to the class of terpenes (limonene, pinene, nerol), for example, are present in much higher quantities in mountain pasture cheeses. What accounts for these variations? Agronomists initially supposed that they resulted from differences in vegetation.

However, in these early studies methods of production and the characteristics of the animals had not been taken into account. Would the observed differences not have been found if the type of cow or the type of fabrication had been modified? In other words, can cheeses typical of mountain pastures be obtained by treating milk from alpine cows with production methods used at lower elevations?

The *Terroirs* of Reblochon

Coulon and his colleagues decided to use Reblochon as a case study. Examining samples taken from several producers who used comparable fabrication techniques, they demonstrated that dietary characteristics determined the sensory characteristics of the cheese. They showed that cows grazing in two different parts of the same mountain pasture—one a southerly slope covered mainly by *Dactylis glomerata* and *Festuca rubra* and the other a northerly slope sparsely planted with *Agrostys vulgaris* and *Nardus stricta* along with unproductive patches of moss and *Carex davalliana*—gave cheeses that differed notably with regard to taste and color. The cheeses from the south-facing precincts were shinier and less yellow, and their taste was more intense, fruitier, and spicier.

The sensory differences between cheeses may be a direct result of the molecular content of the cows' forage. Carotene, which is present in vegetables, gives cheese their color. It is also known that ingestion of certain subtoxic plants (such as *Ranonculus* and *Caltha palustris,* which seem to be more common in north-facing mountain pastures) changes the cellular permeability of mammary tissue and facilitates passage into the milk of enzymes that alter the quality of cheese made from it. Moreover, it is possible that microorganisms typically found in the soil of grazing lands may have a significant influence on the characteristics of cheeses, but this remains to be demonstrated.

The Effects of Silage

Finally, more recent studies have examined the effects of silage, which in certain regions is a matter of controversy (some say it produces mediocre cheese). The new studies involved more than twenty farms producing raw milk Saint-Nectaire of recognized quality. More than sixty cheeses were analyzed by means of sensory studies, with judges instructed to evaluate taste, odor, and texture.

It was discovered that the principal differences between these cheeses resulted from methods of fabrication and the diet of the cows rather than the manner in which their forage was stored. In particular, cows feeding on hay did not always produce cheeses that differed notably from ones made from the milk of cows that were fed on silage. The stocking of forage under controlled conditions therefore has only a limited effect on characteristics other than color.

60

The Tastes of Cheese

Lactic acid and mineral salts give goat cheese its distinctive taste.

AROMAS ARE THE STARS OF THE FOOD INDUSTRY: Many firms produce and sell them to the large food processing conglomerates that make yogurts, soups, sauces, and so on. Nonetheless, foods that are aromatic and little else please only the nose, for they are lacking in taste—hence the interest in taste molecules, still poorly understood. Do these molecules exert the same effect in foods as in water solution, where their properties have long been studied? At the Institut National de la Recherche Agronomique Laboratoire de Recherches sur les Arômes, in Dijon, Christian Salles, Erwan Engel, and Sophie Nicklaus studied this question in connection with goat cheese.

As food is chewed, saliva conveys taste molecules to receptors in the papillae. The Dijon physical chemists sought to analyze the behavior of these molecules in water-soluble compounds, which is to say in the aqueous part of foods. In the case of cheese this phase consists chiefly of lactose (a sugar), lactic acid (formed from lactose by microorganisms in the course of fabrication), mineral salts, amino acids, and peptides (short chains of amino acids). Although the taste of most of these compounds is known, their effects in combination are not. Some of the compounds in the aqueous part of the milk used to make cheese mask the effect of other sapid molecules; others (known as enhancers) augment their effect.

Salles and his colleagues first tested solutions containing only compounds whose presence had been detected in the aqueous phase of goat cheese. Because peptides cannot easily be identified, the chemists isolated them from the hydrosoluble part yielded by 20 kilograms (44 pounds) of cheese. After centrifugation they separated out the juice by a series of ultrafiltrations using membranes permeable by molecules of a mass lower than 10,000, then lower than 1,000, and finally lower than 400. The unfiltered residue was peptides.

All Except One

To evaluate the effect of the various compounds on each of the five basic tastes, a jury of sixteen judges compared the reconstituted aqueous part to a solution with the same components except for one or more compounds that were deliberately omitted. These omission tests were performed under rigorous conditions, with anonymous products, individual booths, red light to prevent bias due to color, and so on. Each taster was equipped with a nose clip in order to eliminate the perception of odors. After training the tasters were instructed to rank each of the five tastes by comparison with a specific reference solution for each sample presented.

Salts, Not Peptides

The first sensory evaluations came as a surprise. Although many studies had offered glimpses of the sapid properties of peptides, suggesting that they have a bitter taste, the peptides in goat cheeses turned out to have no discernible effect on taste, direct or indirect, regardless of their molecular mass. Though not excluding the possibility that these compounds might one day be shown to have a sapid effect in the case of other cheeses, the Dijon team discounted an effect by peptides on the taste of goat cheese.

Comparing solutions containing lactose with ones from which it had been removed, the researchers found that this compound had no effect on the sapidity of the model solutions either. The amino acids likewise turned out to be tasteless. However, lactic acid and mineral salts were found to powerfully contribute to taste. The acidity of the cheese resulted principally from hydrogen ions released by the phosphorus and lactic acids, an effect enhanced by sodium chloride. In the presence of salt, then, the sour note is pronounced. Why? The question has yet to be answered.

The salt taste resulted from the effect of sodium, potassium, calcium, and magnesium chlorides as well as of sodium phosphate. A part of the bitter taste came from calcium and magnesium chlorides, although it was partially masked by sodium chloride mixtures and by phosphates. As for the sweet and umami tastes (the latter caused by monosodium glutamate, widely used in commercial soups and sauces), they were so weak that the researchers were unable to associate them with any of the hydrosoluble compounds tested.

An Overall Taste

The main conclusion to be drawn from these studies, apart from the detailed information they yield regarding the various compounds contained in the hydrosoluble part of the cheeses, is that no taste can be attributed to the action of a single compound. Further complicating matters is the fact that the different sapid compounds have both inhibiting and enhancing effects on one another. On the other hand, we now know which compounds must be added to cheeses in order to reinforce certain tastes or to mask others. Producers may find it difficult to incorporate such compounds, however, not only for legal reasons but also because a large proportion of the molecules dissolved in the milk would be lost during drainage. Nonetheless, gourmets may now amuse themselves by sprinkling their cheeses with various salts and raising toasts. To your chlorides! To your phosphates! To your tartaric acid!

61

Yogurt

A smoother product can be obtained by modifying its milk composition and fabrication process.

HOW DOES ONE MAKE A GOOD YOGURT? The question is poorly posed, for some like their yogurt runny and others like it firm. Ideally what one would want to be able to do, then, is to balance the composition of milk and the method of fabrication in a way that will yield a specific texture and taste. Achieving this objective will take some time, but already Anne Tomas and Denis Paquet of the Danone Group, together with Jean-Louis Courthaudon and Denis Lorient of the École Agro-Alimentaire in Dijon, have shown that the texture of yogurt depends on the microstructure of the milk, which varies according to the concentration of proteins and fats.

The best way to understand the difficulty of the problem will be to make some yogurt ourselves. Put a tablespoon of commercial yogurt in some milk and heat it gently for a few hours. The milk sets—or, as a physical chemist would say, a gel has formed.

Milk is an emulsion, which is to say a dispersion of fat globules and aggregates of casein (protein) molecules in water, in which various molecules such as lactose are dissolved. When one adds yogurt to this emulsion, one is sowing it with bacteria—*Lactobacillus bulgaricus* and *Streptococcus thermophilus*—that transform the lactose into lactic acid. This process of fermentation acidifies the liquid environment and triggers the aggregation of casein micelles in a network that traps the water, fat globules, and microorganisms, which in the meantime have proliferated.

Textures to Order

Unless a great many precautions are taken, yogurt produced by this method is disappointing. A bit of home experimentation will show why. Sowing two identical samples of milk with different yogurts yields different textures and tastes. Similarly, when the milk is curdled with the aid of glucono-delta-lactone, a molecule that progressively acidifies the environment in which it is placed, a still different result is obtained. What is more, causing the milk to curdle at two different temperatures gives different results as well.

Entertaining though they may be, these experiments are not enough to satisfy the needs of the food processing industry, which is obliged to make consistently good products—hence the crucial question posed at the outset. For want of an answer that would put an end to further research, the chemists at Danone and in Dijon restricted their attention to the problem of composition: Given that commercial producers make yogurt from milk that is fortified by the addition of powdered milk, condensed milk, and various milk constituents, how does the composition of this milk determine its microstructure and therefore that of the yogurt made from it?

Because of the variability of these products, the chemists examined emulsions of fixed composition that were prepared by processing a mixture of milk fats and skimmed milk in a microfluidizer (which injects the mixture under pressure into a clear small-diameter tube). Analysis of the light diffused by the various emulsions indicated the size of the fatty droplets.

A Sufficiency of Proteins

Contrary to what prior studies had suggested, the size of the fatty globules did not change with the concentration of fat and protein; what had appeared to be an increase in the size of the droplets, when the proportion of fat is raised, turned out to be only an aggregation of globules of the same size. Naturally the number of droplets grows when the concentration of fatty matter increases, but the proteins are always sufficiently numerous to coat the fatty globules and emulsify them.

Because proteins are not the only molecules that are tensioactive—that is, capable of adhering to the surface of fatty droplets, so that one part is in contact with the fat and the other with the water—the Danone and Dijon chemists studied the changes produced by adding other kinds of tensioactive molecule to emulsions that had already been formed and to mixtures that were subsequently emulsified with the aid of the microfluidizer.

It was expected that molecules with the greatest affinity for fat and water would preferentially attach themselves to the surface of the fatty globules, but experiments showed that this is not the case as long as the tensioactive molecules are put into the mixture before it is emulsified. When the tensioactive molecules are added to an already constituted emulsion, the milk proteins that coat the fatty droplets are not disturbed by these molecules, and their degree of aggregation is unchanged. By contrast, when the tensioactive molecules are introduced at the outset of emulsification, the distribution of the proteins is altered and the degree of aggregation is reduced.

As a result of this research, the prospect of creating new kinds of diet yogurt with the same smooth texture as high-fat ones no longer seems quite so remote.

62

Milk Solids

How to gelatinize milk without destabilizing it.

SLOWLY, OVER CENTURIES, COOKS LEARNED to make solid foods from liquid milk. Cheeses are milk "preserves," made by destabilizing milk and eliminating the water it contains in the form of whey. Yogurt is obtained by heating milk fermented by the bacteria *Lactobacillus bulgaricus* and *Streptococcus thermophilus*. These microorganisms transform the principal sugar in milk, lactose, into lactic acid, which in turn acidifies its liquid environment and causes a network to form throughout the liquid, creating a gel.

In recent years fermentation and curdling methods have been improved, and the texture of yogurt is now known to be determined by the particular procedure used to solidify its milk constituents. This discovery has made it possible to create new milk products. With the aid of gelatinizing and thickening compounds used to make sauces, for example, milk-based desserts have been invented. But unexplained accidents have occurred: When gelatin is added to hot milk, for example, the result often is a disagreeable lumpiness. Jean-Louis Doublier, Sophie Bourriot, and Catherine Garnier, at the Institut National de la Recherche Agronomique station in Nantes, have shown that excessive concentrations of thickening and gelatinizing agents of all kinds have the effect of destabilizing milk.

At first sight this seems a surprising result, considering the varied character of these agents. Gelatin is an extract of animal bones, starches are present in grains and tubers, carrageenans and alginates are derived from algae, galactomannans (guar and carob gums) come from seeds, pectins come from plants, and xanthan gum is obtained from fermented starch.

Destabilizing Sugars

The chemical analysis of these different compounds revealed common features. With the exception of gelatin, all of them are polyosides, compounds of the same chemical family as the sugars; the numerous hydroxyl (–OH) groups found in these molecules are responsible for the thickening of solutions by bonding with water molecules.

But in milk polyosides interact with various dissolved proteins in addition to the casein proteins, which are gathered in bundles called micelles. These bundles are either dispersed in the milk or attached to the surface of the fatty droplets. Why, then, should destabilization be observed when attractive forces seemingly ought to bind the polyosides to the micelles and the fatty droplets?

Doublier and his colleagues explored this question with the aid of laser scanning confocal microscopy, which reveals internal structures without any need for a thinly sliced sample. Using both casein and polyoside labels, they sought to identify areas that were rich in polyosides and proteins, even in mixtures where no destabilization was visible to the naked eye. They discovered that high concentrations of each of the polyosides produce a phase separation: The polyosides come together in certain areas of the solution and the casein proteins in others. In the case of some polyosides, however, this separation sometimes occurs at a very low concentration, although not immediately, because the more viscous the environment the slower the phase separation. Commercial producers who ignore this phenomenon therefore run the risk that their products will separate between the time they are made and the time they are consumed.

Instabilities Arising from a Tendency to Equilibrium

Two Japanese physical chemists, S. Asakura and F. Oozawa, have identified a mechanism known as depletion–flocculation, which occurs in particle suspensions and seems to take place in milk. Particles are in equilibrium when they repel one other (by electrostatic forces between the electrically charged molecules on their surface) more than they attract. Nonetheless, this equilibrium may be disturbed by the presence of a polymer so large that it cannot insert itself between adjacent particles; the polymer concentration is said to be null in this space, which is called a depletion zone. In solution, as a consequence of molecular diffusion, the concentrations of each type of molecule tend to become equal. Because the polymers cannot migrate to the depletion zone, its concentration there is always null, with the result that water leaves this zone in order to reduce the polymer concentration outside it. When the water diffuses in this way the particles are moved closer together. In milk, the casein particles thus form a flocculent mass, forming the dreaded lumps.

This very general phenomenon explains why one finds the same instabilities in milk to which various polyosides have been added. The only way to avoid them is to use as few polyosides as possible. Sailors are fond of saying that a boat can never be too strong; the hull, the rigging, the mast, and the spars should all be reinforced as far as possible in order to avoid rupture. When it comes to food processing, however, this is worthless advice.

63

Sabayons

The foam of a sabayon is stabilized by the coagulation of the egg.

TRY TO IMAGINE WHAT THE PERFECT, ideal, Platonic sabayon would be like. A sabayon (the name is derived from the Italian *zabaglione*) is made by mixing egg yolks and sugar, beating them with a whisk, and adding sweet wine. Heating the mixture while continuing to whisk it, one observes that it becomes foamy, forming a sauce that is served with thinly sliced fruits or sipped as a digestive. The danger, then, is that for one reason or another the preparation won't foam up. Can chemistry and physics be harnessed to guarantee a successful result?

There are many published recipes. Marcelle Auclan, in *La Cuisine* (1951), gives a basic proportion of "1 egg yolk, 1/2 egg shell filled with sugar, two 1/2 egg shells of wine," assuming two yolks per person: "Mix the ingredients together over a very low flame in an enameled saucepan. Stir, stir some more—and keep on stirring. This does not take very long, nor is it at all difficult, but you need to pay attention, and to rely on some sixth sense to know when the sabayon is beginning to set and to become smooth. From the moment—stirring all the while—that you sense a greater density in the liquid, open your eyes and watch very carefully, for it takes only a second for it suddenly to become creamy, all by itself, without having come to a boil."

Mystery lingers, alchemy hovers, and failure threatens. If we use Auclan's proportions as a basis for experiment, translating them first into precise values and then slightly altering them, we can try combining four egg yolks and 200 grams (about 7 ounces) of sugar to make two sabayons. Let's put 30 centiliters (about 10 ounces) of water in one of them but only 10 centiliters (a bit more than 3 ounces) in the other. We will observe that the first one foams, whereas the second one does not.

Water Needed

Why is this so? Keep in mind that egg yolks can be used to thicken sauces and to harden the walls that separate the air bubbles of a foam—as long as the foam forms in the first place. In order for these walls to appear, there must be a sufficient quantity of water. Our second sabayon seems not to have foamed because it didn't have enough water. How can we be sure of this? If we add 20 centiliters (not quite 7 ounces) of water and cook the sabayon a bit longer, whisking as before, we will find that it expands. This is proof that there was not enough water to begin with, but it also provides the cook with a valuable tip: If your sabayon doesn't foam, you can fix the problem by increasing the quantity of liquid.

What role does sugar play in all of this? In addition to contributing to the taste of the dish, its sucrose molecules, which are highly hygroscopic, bind with the water molecules, with the result that the water should no longer be able to form the walls on which a foam's bubbles depend. To test this prediction, compare the first sabayon with two others prepared using four egg yolks, only 100 grams (3 1/2 ounces) of sugar, and 10 and 30 centiliters (roughly 3 and 10 ounces) of water, respectively. The first one rises, but less than the preceding ones, because it contains a lower proportion of water to egg yolk. The second one rises as well, but it reliquifies because it contains too little egg yolk.

Air Bubbles, Not Steam

Let us turn now from composition to cooking. How should a sabayon be heated? Observation, unsupported by theory, seems to show that sabayons expand properly only if the heat is progressively increased. Some chefs believe that steam forms at the bottom of the pan and is trapped there by the coagulation of the egg. If so, one ought to be able to record a temperature of 100°C (212°F) at the bottom of the pan when the sabayon begins to foam. However, other chefs believe that a sabayon has to be brought to a temperature of only 80°C (176°F) in order to foam. Which assumption is correct?

Let's try to reason our way to an answer. The point of using a whisk to prepare a sabayon, obviously, is to introduce air into the sauce. And the only physicochemical transformation that occurs during cooking is the coagulation of the egg yolks, which begins at 68°C (154°F). Therefore one would expect that there is no need to heat the sabayon to the boiling point, or even to 80°C (176°F), in order for it to foam (the white of an egg foams at room temperature) and that a temperature of 68°C (154°F) is enough to stabilize the foam by rigidifying the walls that separate the air bubbles through the coagulation of the proteins in the egg yolk.

Experiment confirms these expectations. The question remains which of these three cooking temperatures is to be preferred. Leading pastry chefs agree that a perfect sabayon should not have the taste of cooked egg. For this reason a sabayon cooked at the lowest temperature would be the best. True, it will take a bit longer to make, but what gourmet would not be willing to wait for a Platonic sabayon?

64

Fruits in Syrup

How to optimize the sugar concentration of syrups for preserving fruits.

AUTUMN DRAWS NEAR, and with the coming of cold weather summer's fruits will soon vanish. How can they best be preserved to last through the winter? Several methods are common, especially canning and freezing. In each case one seeks to prevent the proliferation of germs, which at room temperature occurs rapidly in foods that contain sufficient quantities of water and organic matter to favor the development of microorganisms.

Anyone who has tried to preserve fruit in syrup is familiar with the need to steer a safe course between the Charybdis of fruit that swells to the point of falling apart and the Scylla of fruit that shrivels up. The observant cook will suspect that the first danger is encountered when the syrup contains too little sugar, and the second presents itself when the syrup has too much sugar. What does one need to know to get the amount of sugar right?

Experience and Science

Answering this question involves analyzing a phenomenon that occurs during cooking. Cookbooks generally recommend using fruit that has not completely ripened. Once the pieces have been carefully pierced (use a needle or the tip of a sharp knife), they are to be put in a "20° syrup" and heated in glass jars packed tightly together with cloth and covered by boiling water. The cooking time is said to depend on the size of the fruit: only two minutes for currants, for example, but five for apricots.

The reason for doing this is to destroy any microorganisms that are present by heating, a technique that has been widely used since the earliest days of canning. But cooks today deserve more than simply being told to follow the instructions of an old recipe. They ought to be told, in particular, that the degree to which microorganisms are eliminated is proportional to time and temperature. In this case, with the temperature fixed, time is what determines the degree of sterilization. Classic recipes therefore are mistaken in prescribing different times depending on the size of the fruits because all types of fruit must be cooked for the same length of time to obtain the same degree of sterilization. But this means that the smallest fruits are in danger of being overcooked, which will not come as welcome news to gourmets.

On the other hand, it is experience rather than science that explains the use of cloth in packing the glass jars together. When the jars are heated in a preserving pan, the boiling of the water violently shakes the jars, which are likely to shatter if they are not protected.

The Strength of Syrups

Now we come to the problem of the syrup. Recipes often call for a syrup of 20°, without further explanation. The degrees in this case obviously are not degrees Celsius or Fahrenheit, but one has to know something about the art of confectionery to realize that there is a possible confusion between degrees Baumé (equal to $145 - 145/S$, where S is the specific mass of the sugar) and degrees Brix (or Balling), which represent the proportion of sugar to mass. Why don't cookbook authors simply speak of the quantity of sugar to be dissolved per liter of water? In the case of fruits in syrup, as we will see, the cook can safely ignore such complications.

The main problem in preserving fruits in syrup, assuming the sterilization process has been carried out properly, is that they swell up in syrups that are too light and shrivel up in syrups that are too concentrated. What accounts for these phenomena? Fruits are modified by osmosis: Preserved for a sufficiently long time in syrup, they evolve toward equilibrium, which is to say a balance between the concentrations of water in the fruit and in the syrup.

We can verify this by looking in a microscope: A plant cell in the presence of a sugar crystal is emptied of its water, which diffuses through the cell wall and membrane, equalizing the concentrations of water inside and outside the cell. Given that a fruit is itself an assembly of cells, the same phenomenon occurs on a macroscopic scale in canning.

If the syrup is light, the fruit—having a lower concentration of water—absorbs the liquid outside it, expands, and finally bursts. Conversely, a syrup that is too concentrated draws out the water inside the cells. One observes the same phenomenon when gherkins and eggplants "sweat" in the presence of salt, for example.

Nonetheless, this explanation does not tell us how to gauge the right strength for syrup. Let's prepare syrups using 10 grams of sugar per liter, 20 grams per liter, and so on (roughly 1/3 ounce per quart, 2/3 ounce, etc.), until the saturation point is reached, and put fruit in each of them (plums, for example, pierced as the recipe indicates). One observes first of all that the fruit floats in the syrups containing the most sugar but not in the others. This is not surprising, for the density of a syrup increases with the amount of sugar dissolved in it. If we wait to see the effects of the sugar concentration we will find that our patience is rewarded, for it becomes plain that in heavy syrups, in which the fruit floats, it has shriveled up, whereas in light syrups, in which the fruit has sunk to the bottom, it has broken apart.

Of course, the correspondence is not exact, for many kinds of fruit exhibit varying degrees of ripeness. Some pieces float; others fall to the bottom of the pan because their composition and, in particular, their sugar concentration has changed in the course of ripening. Moreover, the pit of the fruit may not have the same density as the surrounding flesh. We can complicate the experiment somewhat by making a series of increasingly concentrated syrups and putting into them whole fruits, quartered fruits with the pit still attached, quartered fruits separated from the pit, and pits by themselves. The density of the pit turns out to be irrelevant, for the variability in ripeness of the individual pieces of fruit is greater than the difference in density between the flesh and the pit.

An efficient way to determine the right strength of the syrup is this: Make the syrup a little too strong, so that the fruit floats on top, and then add some water. When the fruit ceases to float, the desired concentration has been achieved. Let winter come!

65

Fibers and Jams

Pectins can be recovered from the fibrous matter of fruits and vegetables by extrusion cooking.

DURING THE WINTER, THE SPECTER OF A JAM that fails to set haunts cooks who make orange marmalade. Sometimes, instead of forming the expected semisolid mass, the juice remains juice. How can this culinary debacle be avoided? Jean-François Thibault, Catherine Renard, Monique Axelos, and Marie-Christine Ralet at the Institut National de la Recherche Agronomique station in Nantes have found an answer to this question, although the purpose of their research was mainly commercial. They showed that the technique of extrusion cooking, used especially in fabricating cocktail crackers, makes it possible to extract large quantities of pectin, which is to say gelatinizing molecules, from the pulp of lemons, oranges, and beets.

The occasional failure of jams to set seems paradoxical. Pectins are molecules that, together with cellulose and other polysaccharides (long molecules formed by the linking of simple sugars), form the cell walls of most plants; only cereals do not contain them. But because the method of moderate heating used to prevent volatile aromas from evaporating does not efficiently break down the cell walls, the pectins are unable to come together to create a network, or gel, that fills up the entire volume of the solution.

In the case of commercial production a different problem arises: The residue from pressing apples to make apple juice and cider, the pulp of lemons and oranges left over after juicing, and the pulp of beets that remains after their sugar has been extracted all contain large quantities of plant fiber that producers would like to find a use for. Traditionally pectins are recovered from these fibers through heating in an acid solution. In this way the pectins are detached and then, in a further step, purified. The resulting material is used as a gelatinizing or thickening agent or to coat potato chips. In the future it may find a new role as a detergent for cleansing polluted water (because pectins bond strongly with the metallic ions that are the source of water impurities).

Nonetheless, the acid extraction process has several disadvantages. In particular, it threatens to degrade pectin molecules by breaking the bonds between their constituent sugars and altering the chemical groups carried by these sugars that are responsible for gelatinization.

From Polymers to Dietary Polymers

The Nantes chemists had studied another method known as extrusion cooking for several years in connection with the processing of meats and starches, which gave them the idea of using it to extract pectins from fruit pulp. The apparatus for this technique (adapted from devices used to extract polymers) has one or more Archimedean screws that rotate at variable speeds, shearing off the fibrous material and injecting it into a tube, where it undergoes rapid expansion. The extracted material then passes through a series of chambers in which it is heated at controlled temperatures.

Not all pectins promote gelatinization. Their properties depend on their chemical composition, which varies according to the fruit or vegetable from which it is derived: Whereas pectins from limes, lemons, oranges, and apples are efficient agents, the ones extracted from beet pulp do not gelatinize (although beet pectins, because they chelate heavy ions, produce superabsorbent gels after chemical modification). The differences in the properties of pectins are a consequence mainly of esterification. The sugar chains that form their molecular skeleton carry carboxylic acid ($-COOH$) groups, which are esterified to varying degrees according to the fruit. As a result, these groups now display the form $-COOCH_3$. Depending on the extent of esterification, the pectins associate more or less easily with one another. It is this ease of association that ensures gelatinization. Extraction for the purpose of producing gelatinizing or thickening agents therefore must respect the integrity of the pectin lateral chains.

The extrusion cooking method has several advantages. Not only is the extractive apparatus much simpler and cheaper than that of other techniques (the screws used at Nantes are only a meter long), but the processing is rapid and can be automated. Additionally, it yields as many pectins as the traditional acid treatment and preserves their molecular structure. The Nantes researchers found that the shearing off of plant matter during extrusion cooking is the critical step. Indeed, if this material is not heated, the molecular integrity of the pectins is preserved despite compression.

Culinary Moral

How can you take advantage of these results cooking at home? At room temperature, squeeze the juice from the fruit and set it aside, finely grind up the remaining fibrous matter in a food processor, then add the juice back to the ground-up material and cook over low heat. But because the pectin molecules are apt to bond with the copper atoms of your pan instead of with one another, don't leave your jam to cool in the pan. Let it set in glass containers instead.

66

The Whitening of Chocolate

To keep chocolate from turning white, keep it chilled.

WHY DOES CHOCOLATE BECOME COVERED with an ugly white film after a few days? Michel Ollivon and Gérard Keller of the Centre National de la Recherche Scientifique (CRNS), in collaboration with Christophe Loisel and Guy Lecq of the Danone Group, have recently explained how certain constituents of this partially liquid mixture migrate and crystallize on its surface, causing it to change color.

Chocolate is a dispersion of sugar crystals and cocoa powder in cocoa butter. Once crystallized, the cocoa butter serves as a binding agent for the solid particles, just as cement binds the sand and gravel in concrete. Nonetheless, chocolate does not easily cohere: The sugar crystals are hydrophilic, whereas the cocoa butter is hydrophobic. Master chocolate makers add lecithin molecules as part of a process known as conching in order to promote the coating of the sugar by the cocoa butter.

Since the 1960s it had been known that the whitening of chocolate resulted from the thermal behavior of cocoa butter. The mixture of semisolid and semiliquid fats that make up cocoa butter can crystallize in six different ways (called forms 1–6), and only a precise sequence of reheatings and recoolings, collectively known as tempering, yields crystals that are sufficiently stable to prevent chocolate from whitening. The recooling phases are crucial; failure in any one of them will ensure a disappointing result.

Chocolate X-Rayed

The CRNS and Danone researchers investigated the crystallization process by measuring variations in viscosity during tempering. X-rays showed that well-tempered chocolate is crystalized in form 5, and further observation by means of polarizing microscopy disclosed that the surface of the chocolate immediately after fabrication is free of crystals; only a few holes, no doubt created by the bursting of air bubbles on the surface, and a few minute cracks are visible. By contrast, when the chocolate is subjected to substantial variations in temperature, the whitening appears after a few days. Both the surface and the interior crystalize in form 6.

How are we to interpret these observations? It was long thought that the whitening resulted from a migration of a part of the cocoa butter toward the surface, where it recrystallized, probably in form 6. Today we know that the composition of fatty matter in the body of the chocolate differs from that of the whitened surface. Hervé Adenier and Henri Chaveron at the University of Compiègne have confirmed that form 6 crystals are more stable than those of form 5; the difference in their fusion temperatures is 1.5°C (2.7°F). Moreover, form 6 is more compact than form 5. During the transition between the two forms, a part of the cocoa butter is pushed out toward the surface of the chocolate.

The difference in fusion temperatures corresponds to a difference in composition. The molecules of cocoa butter, triglycerides, are like a comb with three teeth: The base of the comb is a glycerol molecule, and the three teeth are fatty acid molecules. Carbon atoms in these acids are held together by means of single or double bonds. The fusion temperature of a triglyceride diminishes with the number of double bonds.

Pushed by Fusion

The proportion of the liquid phase naturally increases with temperature while becoming enriched by triglycerides having one double bond, which melt at a lower temperature. Whitening corresponds to the recrystallization, during cooling, of these enriched liquid parts.

Whitening threatens mainly chocolates whose interior, or lining, is rich in fats. When two blocks of fatty matter are combined, their fats become mixed together, especially at high temperatures (when the proportion of liquid and molecular motion both increase). The surface crystals of the whitened chocolate are quite different from those of the interior mass, being composed for the most part of triglycerides having a single double bond.

The composition of the whitened crystals is virtually invariable, no matter what the fatty content of the lining, and similar to those of whitened chocolate without a lining. The migration of fats in the lining toward the covering layer of chocolate leads to an enlargement of the liquid phase of the covering layer. Consistent with the findings of the CNRS and Danone researchers, this enlargement accelerates whitening without changing the composition of the outer crystals.

The moral of the tale is that if you want to preserve the aroma and appearance of chocolate, store it at a low temperature, say 14°C (57°F). In this way you will limit the migrations that accompany storage. Then warm up the chocolate before eating it.

67
Caramel

The molecules of caramelization finally identified.

SENECA MENTIONED CARAMEL as early as 65 BC, but for more than 2000 years the details of the chemical reactions that give heated sugar its inimitable flavor were unknown. Exploiting recent results in the chemistry of sugars and using modern analysis techniques, Jacques Defaye and José Manuel Garcia Fernandez at the Centre National de la Recherche Scientifique laboratory in Grenoble recently elucidated the structure and mechanisms responsible for the formation of the odorant and taste molecules that make up caramel.

Along with the Maillard reaction, which generates the aroma of roast beef, coffee beans, beers, and bread crust, caramelization is one of the principal methods for the culinary transformation of foods that contain sugars. Whereas the Maillard reaction is a reaction of sugars with amino acids or proteins, caramelization involves only sugars. It is probable that the two reactions jointly play a role in the cooking of most foods containing sugars, the share of each depending on the relative quantities of sugars and proteins.

Although caramelization has influenced the taste and appearance of dishes ever since sugars were first heated, exactly how these transformations take place remains a mystery, and an economically important one at that: In France alone the food processing industry produces 15,000 tons of caramel per year, which are used in the making of milk, cookies, syrups, alcoholic beverages, coffee, and soups.

A Scientific Tradition

The first scientific studies of caramel were done in 1838 by the French chemist Étienne Péligot. For the next twenty years caramel was consigned to purgatory, until M. A. Gélis, Charles Gerhardt, and Gerardus Johannes Mulder proposed in 1858 to divide its nonvolatile component (making up 95% of the caramelized product) into three parts: caramelan, caramelene, and caramelin. Nonetheless, these substances, obtained from successive dissolutions with alcohol and water, were no more clearly defined, chemically speaking, than the famous osmazome that Thenard and Brillat-Savarin claimed to constitute the sapid principle of meats. None of the parts extracted by precipitation is constituted by a single type of molecule.

Investigation resumed in the early twentieth century. Caramel was then believed to contain humic acids, poorly understood reducing compounds whose tanning properties are also found in lignite. The various compounds of the volatile part of caramel were also discovered, including 5-hydroxymethyl-2-furaldehyde and some twenty other compounds that contribute to its penetrating odor (including formaldehyde, acetaldehyde, methanol, ethyl lactate, and maltol).

Subsequently it was observed that caramelan reacts with alcohols. Analysis of the nonvolatile part nonetheless remained a nagging problem until 1989, when modern research methods made it possible to detect the presence of a derivative of glucose.

Water Eliminated

Sucrose is a disaccharide composed of glucose–fructose bonds. Each of these two subunits has a skeleton composed of six carbon atoms. Five of these atoms each carry a hydroxyl (–OH) group. The sixth one bears an oxygen atom attached by a double bond, with a glycosidic bond such as $-CH_2-O-CH_2-$ binding the two rings. Applying the same methods of analysis they had used in studying the chemistry of sugars, the Grenoble researchers elucidated the main features of the chemical transformations of the nonvolatile part of caramel. Among other things they observed the formation of fructose dianhydrides, in which two fructose rings are connected by two $-CH_2-O-$ bonds, which in turn define a third ring lying between them. Several molecules correspond to this description because sugars come in many isomeric forms, which is to say that molecules having the same atoms can differ if the atoms are linked by different bonds.

Finally, the Grenoble chemists showed that during the caramelization of sucrose, for example, the nonvolatile part results from an initial reaction dissociating the sucrose into glucose and fructose. These elementary sugars then recombine, forming oligosaccharides having various numbers of elementary sugars: The glucose may combine with glucose or fructose, the fructose may react with fructose, and so on.

These recent results are commercially important, for they make it possible to consider polydextroses—used to give texture to dishes in which sugar is replaced by intense sweeteners—as naturally occurring compounds. Because polydextroses are naturally present in caramel, they are not subject to the same system of regulation as other synthetic molecules. Moreover, the tendency of various glucides to caramelize can now be investigated more easily.

68

Bread and Crackers

The mechanical behavior of bread resembles that of plastic materials.

LEFT OUT IN THE KITCHEN, at room temperature, bread goes stale. Frozen, it seems to change more slowly, but at what temperature must it be kept in order to stay in the same state as when it comes out of the oven? 7°C (45°F)? 0°C (32°F)? –10°C (14°F)? Physical chemists at the École Nationale Supérieure de Biologie Appliquée à la Nutrition et à l'Alimentation (ENSBANA) in Dijon have sought to answer this question using their knowledge of polymers, which are very long molecules formed by the linking of subunits called monomers. This seemed to be a natural approach, for foods contain many polymers: The molecules that constitute the starch granules in flour are linear or ramified chains of glucose molecules known respectively as amylose and amylopectin, proteins are chains of amino acids, and so on.

At high temperatures polymers are in a liquid state because they have sufficient energy to move in a disordered fashion, allowing their mass to flow. When polymers are cooled, they initially form a rubbery solid in which certain polymer chains crystallize while preserving the ability to slide past one another. Then, at temperatures lower than the temperature of vitreous transition, the chains are immobilized and the material solidifies, with their crystalline parts dispersed in an amorphous rigid part, or glass. The structure of the solid phase depends on the cooling. When the cooling is rapid, the viscosity increases too quickly for the molecules to be able to crystallize, and the vitreous part predominates.

Thus many foods are kinds of glass: Sugar cooked with water becomes concentrated with the evaporation of the water and gradually forms a glass; powdered milk, coffee, and fruit juice sometimes also appear in a vitreous state. What about a fresh loaf of bread? Is it initially a rubbery solid that then vitrifies or partially crystallizes as it goes stale? Martine Le Meste, Sylvie Davidou, and Isabelle Fontanet at ENSBANA studied this question by recording the mechanical behavior of various hydrate samples as a function of temperature and comparing the reactions of loaves of bread with those of extruded flat breads, such as crackers.

When one heats bread dough, which is essentially a mixture of flour and water, the starch granules in the flour release their amylose molecules into the water, as we have seen. As the bread cools, the amylose molecules form a gel that traps the water and the amylopectin. In order to prepare variously hydrated breads, the Dijon team first completely dehydrated a series of samples by placing them for a week in desiccators, where the water was absorbed by phosphoric anhydride. The samples were then rehydrated under controlled hygrometric conditions and coated with an impermeable silicone grease. A viscoelastometer was used to measure the force transmitted by the samples when they were deformed in a controlled way, yielding a coefficient of rigidity known as Young's modulus.

The researchers found that bread remains in a rubbery state as long as the temperature is higher than the vitreous transition temperature, –20°C (–4°F). On the other hand, analysis of the vitreous transition temperature as a function of water content showed that a part of the water does not freeze and that it plays a plasticizing role.

Freezing Bread

These observations have practical implications. The many results obtained by polymer chemists allow us to predict the changes in the mechanical properties of bread and its cousins as a function of their water content, crystallinity, and so on. Among other things, even if the water that freezes is immobilized, freezing will not arrest such changes as long as the temperature is higher than the vitreous transition temperature. At temperatures between –20°C and 0°C (–4°F and 32°F), then, bread continues to undergo structural alteration. To preserve bread without compromising its textural characteristics, the freezing temperature must be lower than the vitreous transition temperature.

The loss of freshness in bread had long been attributed to the phenomenon of starch retrogradation, in which amylose progressively crystallizes, releasing its water. The Dijon team observed instead a co-crystallization of amylose and amylopectin into hydrated crystals. Lipids counteract the loss of freshness that occurs over time because they bind with the amylose, forming crystals that retard the co-crystallization of the amylose and amylopectin.

Nonetheless, the firmness associated with stale bread does not result solely from this co-crystallization. The behavior of the amorphous, or vitreous, regions seems to play a major role. Water is an important parameter in storage, for it works to plasticize these regions, which in turn affects the rate and type of crystallization that occur.

69

The *Terroirs* of Alsace

The openness of the landscape is a crucial factor in winemaking.

THE WORLD OF WINE AND VINE so little doubts the existence of differences in the overall natural environment—the *terroir,* as it is called—of winegrowing regions that it has made them the basis for awarding protected designations of origin. Is this justified? Agronomists are accustomed to examining how the particular features of a given viticultural site—its climate, soils, and parent rock—affect the growth of its vines. Éric Lebon and his colleagues at the Institut National de la Recherche Agronomique (INRA) station in Colmar have studied defined sections of the Alsatian landscape and shown that its openness is at least as important as its capacity for retaining groundwater and exposure to sunlight.

Wine growers seek to plant grapevines in conditions that favor the formation of berries, rather than leaves or branches, and the accumulation of sugars (for fermentation) and aromas. Berries are able to ripen before the intemperate weather of autumn causes them to rot only if the vines have begun to grow early enough. For this reason it was long thought that sunshine was the chief advantage of good *terroirs.*

At the request of the Centre Interprofessionel des Vins d'Alsace, the Colmar agronomists continued research that had been begun in the 1970s in the Bordeaux region by Gérard Seguin and his colleagues at the Institut d'Œnologie there. The team led by Seguin analyzed the importance of the soil, and the way in which it nourishes the vine with water, in promoting the vine's growth. The best *terroirs,* it found, enjoyed a regular supply of water and periods of only moderate drought, conditions that encourage the early ripening of the grapes.

Beginning in 1975, René Morlat and his colleagues at the INRA station in Angers studied the land in the legally protected red wine–producing vineyards of the Loire Valley (Cabernet Franc from Saumur-Champigny, Chinon, and Bourgueil). Their work confirmed the earlier observations and showed that the more rapidly the soil heats up in the spring, the earlier the vine develops and the more favorable the landscape is to successful cultivation. The Angers agronomists suspected that the relevant climatic characteristics could be analyzed in terms of a landscape's openness, determined by measuring the angle of the horizon in relation to a horizontal axis for the eight principal directions of the compass dial. Vines were supposed to develop differently in a basin, where the degree of openness is low, than at the top of a hill, where the degree of openness is high. Lebon and his colleagues tested this assumption in the vineyards of Alsace, which are semicontinental (and so quite different from those of Bordeaux and the Loire Valley), in an area lying between Witzenheim and Sigolsheim, near Colmar.

The soils and parent rock of Alsace had already been extensively studied but never on the scale of the experiment now contemplated. Over an area of 1,750 hectares (4,325 acres) Lebon and his colleagues carried out a hectare-by-hectare survey to determine homogeneous pedological units. On the basis of various criteria for characterizing types of soils, they distinguished some thirty homogenous units. A pit was dug in each of these zones to analyze soil types.

Landscape and Climate

Lebon and his colleagues made climatological measurements at stations set up near pits dug in vineyards in which the Gewurtztraminer grape was cultivated and observed that local climates varied little, on an annual average basis, from unit to unit. Significant variations occurred only on shorter time scales. During periods of inclement weather, for example, the air temperature was found to depend only on altitude; in clement weather the diurnal temperature varied as a function of altitude, declivity, orientation, the height of the horizon from east to west, and the thermal properties of the soils. Taken together these measurements define mesoclimates, which is to say climates on the local scale of a slope or valley bottom, for example. Comparing the various landscape indices with the climatological data made it possible to refine the notion of *terroir*. Temperature turned out to be important in determining the mesoclimate, but not as important as the landscape.

What makes a *terroir* good for growing wine? The INRA agronomists showed that in Alsace, more than in the Bordeaux region or the Loire Valley, the principal differences result from variations in the maturity of the grapes at the moment of harvest. In all the wine-growing regions studied, water nourishment conditions played a major role in determining, among other things, the length of time between the fruiting of the vine and the ripening of the grapes. Maturation comes late when water is plentiful because the vine produces leaves rather than berries. When the supply of water is insufficient, maturation is delayed as well, but the adage that the vine must be made to suffer if it is to produce grapes is only partly true: The supply of water should decline both moderately and regularly.

Now that the notion of *terroir* has been validated in Alsace, it remains to study the relationship between a wine-growing site's total natural environment and the quality of the wines it produces. Attempts to characterize aromas are under way: Alex Schaeffer and his colleagues at the INRA station in Colmar have already observed a high degree of variability in terpene alcohol content and the oxides of these alcohols.

70

Length in the Mouth

Enzymes in saliva amplify an important aromatic component of wines made from the Sauvignon Blanc grape.

BIOCHEMISTS ARE VERY INTERESTED in the methods used for making wines, but they have rarely explored the physiology involved in tasting them. Recent results suggest that this situation may be changing. In wines made from the Sauvignon Blanc grape an odorant molecule has been found whose effect is registered only when the enzymes in saliva have separated it from its precursor. A few moments are needed, then, for the aroma to be perceived.

In 1995, Philippe Darriet and Denis Dubourdieu at the Institut d'Œnologie de Bordeaux discovered a molecule that gives Sauvignon wines a boxwood or broom note. Significantly, this simple molecule, whose skeleton is composed of only five carbon atoms, contains a sulfur atom. Darriet and Dubourdieu, observing that it appears during alcoholic fermentation, managed to identify its precursor. Additionally, they observed that the frequency with which this precursor is transformed into an odorant molecule depends on the strains of yeast responsible for fermentation. The Bordeaux chemists studied a class of enzymes capable of releasing the wine's distinctive aroma, reasoning that if they could identify the particular enzyme at work in the case of Sauvignon they would have thereby determined the structure of the precursor.

In a related study, Claude Bayonove at the Institut National de la Recherche Agronomique station in Montpellier examined glycosidases, enzymes that dissociate odorant molecules of the terpene class from the sugar molecule to which they are bound. The molecules considered in the Montpellier study do not produce the odorant compound analyzed in connection with the Sauvignon grape because this compound is not bound to a sugar.

Amino Acids and Sulfur-Based Aromas

Darriet and Dubourdieu looked at enzymes that break the bonds between a carbon atom and a sulfur atom to see whether one of them causes Sauvignon's aroma to appear. They were interested in an enzyme called lyase, produced by the intestinal bacterium *Eubacterium limosum,* that breaks down the sulfur derivatives of cysteine, an amino acid. The compound in Sauvignon was released in vitro from a nonvolatile extract of its must through the action of some product present in the medium obtained by grinding up *Eubacterium limosum.* They concluded that the precursor of the aroma in the must contains cysteine.

These results confirmed, first, that the Sauvignon grape has an aromatic potential that is brought out by the vinification process. This means that the wine maker must choose yeasts that are capable of developing the taste latent in the grape. The great gastronome Curnonsky (1872–1956) was famous for demanding that foods have "the taste of what they are." Chemistry is a way of satisfying this requirement.

The Origin of Length

The Bordeaux chemists also showed that the aroma's precursor is found in significant quantities in the wine itself, the aromatic molecule remaining bound to the nonaromatic part. The operation of enzymes in the saliva, which during tasting detach the cysteine from the sulfur aroma after a few seconds, is the chief mechanism underlying the notion of length, which is to say the length of time the sensation produced by a wine remains in one's mouth. A special unit, the *caudalie,* has even been introduced to quantify the length of time this sensation persists once the wine has been swallowed. Certain Bordeaux wines produce a sensation that lasts several seconds (or *caudalies*) and, in a few exceptional cases, is capable of reasserting itself: Having disappeared, the sensation comes back. If the sulfur molecule discovered by the Bordeaux researchers does not actually explain this peacock-tail effect, it is the first to yield a length whose origin is understood.

What are the practical consequences of this discovery? Because the adulteration of wines is forbidden by law, oenologists (who often work for producers) will seek to optimize the concentration of a given aroma during vinification. But if you don't have a lot of money and are willing to indulge in a little amateur chemistry, you can try adding the aroma to wines in your collection that don't have enough of it. After all, the law doesn't forbid lengthening the duration of one's pleasure.

71

Tannins

The development of tannins diminishes the astringency of wines.

HOW DOES WINE AGE? Gourmets have long complained of the lack of scientific interest in the role of tannins in the development of red wines. Tannins are astringent substances that are abundant in young wines and that change as the wine ages, giving it an adobe color, smooth taste, and powerful aroma. Over time, tannins are said to soften up. With the aid of modern analysis techniques, Joseph Vercauteren and Laurence Balas at the University of Bordeaux revived a topic of research that had largely been abandoned since the work of Yves Glories in 1976 and elucidated several chemical transformations whose existence had previously been suspected by oenologists.

Tannins are what create the sensation of astringency when one chews rose petals, for example, or drinks a wine that is too young. By forming complexes with the lubricating proteins in saliva they prevent these proteins from playing their natural role, so that the mouth feels dry. Tannins are found in the woody parts of plants as well as fruits. The alcoholic solution that is formed in the process of making red wines, in particular, extracts tannins from the pip, skin, and stalk of the grape.

Thirteen years after Glories wrote his thesis on the "coloring matter of wines," in 1989, Vercauteren and Balas analyzed the behavior of tannins in the course of vinification. Every other day they took samples from two red wines (Château Cérons and Château Baron Philippe de Mouton Rothschild) in order to monitor variations in tannic concentration. Using ethyl acetate to extract the tannins from these wines, they obtained paradoxical results. Whereas certain oenologists believed that maceration must be prolonged beyond two weeks in order for tannins to accumulate, giving the wine body and a chewy sensation, the Bordeaux chemists discovered that the mass of recovered tannins for the three grapes tested (Merlot, Cabernet Sauvignon, and Cabernet Franc) was greatest around the tenth day of vinification. Why this contradiction? Was it because tannins, being reactive molecules, are transformed into compounds that are difficult to extract when they are put into an ethyl acetate solution?

This initial finding made it clear that a fresh start had to be made, synthesizing the constituent elements of tannins and studying their chemistry in the hope of being able to identify them in the wines. Vercauteren and Balas therefore sought to develop a method for hemisynthesizing condensed tannins, which is to say for creating flavonol-based compounds derived from the parent ring structure of flavan that contain several hydroxyl (–OH) groups. Taxifoliol, extracted from the bark of the Douglas fir, contains several preponderant compounds whose chemical structure was subjected to analysis.

The question needing to be answered, then, was how flavonols bond with condensed tannins. The simple case of flavonolic dimers, formed from the combination of only two flavonols, was already difficult enough, for its two subunits can bond in two different ways. The Bordeaux chemists compared the structure of the synthesized compounds and various natural tannins by nuclear magnetic resonance imaging, but this method did not disclose the two types of bonds. By contrast, transforming the tannins into peracetates by replacing –OH hydroxyl groups with $-OOCCH_3$ acetate groups yielded a general method for determining the structure of condensed tannins.

Armed with this result, Vercauteren and Balas sought next to determine whether flavonols and glycosylated tannins (tannins bonded with sugars) were present in the wines they had selected for examination. They knew that the glycosylation of certain hydroxyl functions stabilizes polyphenols (the class of molecules to which tannins belong), but condensed tannins had never been observed in glycosylated form in either grapes or wine. Nonetheless, they suspected that the diminishing astringency of wines over time resulted from an intensification of tannin–sugar bonds. Resorting once again to a method that had served them well in the past, they synthesized glycosylated tannins, analyzed their characteristics, and then sought to identify these molecules in the wines.

Because the compounds extracted by means of ethyl acetate were rich in two simple flavonols, catechin and epicatechin, the chemists studied its glycosylation using glucose, the sugar found in grape juice. Four glycosylated flavonols were shown to be highly insoluble in ethyl acetate, which corroborated the initial hypothesis: The reason that flavonol and glycosylated tannins had not been found in either the grape or the wine was that the extractive solvents were unsuitable to the task. Subsequent analysis of these hard-to-extract compounds indicated that at least three glycosylated tannins are present in both the grape and the wine, thus establishing that the polymerization and glycosylation of tannins are two of the mechanisms responsible for the aging of wines.

72

Yellow Wine

Sotolon is the principal molecule that gives *vin jaune* its characteristic flavor.

IN 1991, Patrick Étiévant and Bruno Martin at the Institut National de la Recherche Agronomique (INRA) station in Dijon began to analyze a wine known as *vin jaune* (yellow wine) that is produced only in the French Jura. The specific flavor of this wine results from the practice of maturing it in barrels for several years under a thick veil of yeast of the species *Saccharomyces cerevisiae*. A similar wine is made in Alsace, Burgundy, and in the town of Gaillac, in the Tarn, where it goes by the name *vin de fleur* or *vin de voile*. Its only near equivalents outside France are the sherries of Spain and the Hungarian tokay. The Dijon chemists wanted to know which molecules are responsible for its distinctive flavor.

This wine and ones like it contain hundreds of volatile components, a tenth of them aromatic, so that identifying the molecule that produces a particular aroma is not easy—a bit like picking out a guilty person among some 300 suspects. In the early 1970s some researchers believed that solerone (4-acetyl-gamma-butyrolactone) was the chief odorant molecule of yellow wines, but in 1982 Pierre Dubois, also at the Dijon station, found it in red wines as well. Solerone had an alibi.

The next suspect was sotolon (4.5-dimethyl-3-hydroxy-2[5H]-furanone), a molecule constructed around a ring of four carbon atoms and one oxygen atom. Because sotolon and solerone are found in minimal concentrations in *vins de voile* and are chemically unstable, the Dijon chemists searched for ways to extract them more efficiently.

Sotolon Uncovered

The most direct method of analyzing molecular extracts from wine is chromatography. One begins by injecting a sample into a solvent, which is then vaporized. Next, a polymer-lined tube is inserted that captures the various compounds of the gaseous mixture in varying proportions, with the separated compounds settling at the bottom of the tube. The chemists' first task was to devise a variant of this technique in order to identify the compounds present in minimal quantities in the complex mixtures of yellow wine.

Chromatograms of the wine samples were then compared with those of pure solutions of sotolon and synthetic solerone. Sotolon was found to be present in 40–150 parts per thousand in sherries. Solerone seems to be less characteristic of yellow wine, but its concentrations are higher in sherries, which explains why it was first found in these wines. Finally, sensory tests of the separated parts showed that tasters did not perceive solerone, in the concentrations in which it is found in Savagnin (the grape from which yellow wine is made), either in the wines themselves or in the laboratory solutions. Solerone therefore was unquestionably not the culprit in the case of the *goût de jaune.*

Beginning in 1992, the chemists devoted all their energies to the search for sotolon, whose presence had been observed in cane sugar molasses, fenugreek seeds, soya sauce, sake, and other substances. It was also known to be present in certain wines made from overripe grapes attacked by the botrytis fungus (*Botrytis cinerea*), better known as the noble rot. This fungus is responsible for the distinctive character of Sauternes, for example, and so-called late harvest wines. Sotolon was not found in either red wines or oxidized wines. Moreover, its perception threshold was determined to be only 15 parts per thousand.

Tasting tests found the typical character of *vins de voile* to be most pronounced, exhibiting a note of walnut, when the sotolon concentration in these wines is high. In even greater concentrations the tasters detected a note of curry.

The Death of Yeast

The sotolon trail was then taken up in 1995 by another INRA researcher in Dijon, Élisabeth Guichard, who developed a method for rapidly measuring its concentration. In *vin de paille* (or straw wine, a sweet white wine made from grapes dried on straw mats), this was found to be 6–15 parts per thousand.

In yellow wines, sotolon is synthesized at the end of the yeast's exponential growth phase: In vintages that have been aged in casks for one, two, three, four, five, and six years, respectively, the quantity of sotolon is small in the early stages of maturation and rises notably after four years, especially in cellars that are not too cool.

Samples taken at different depths beneath the yeast veil revealed that sotolon is twice as concentrated in the middle and at the bottom of the casks as it is just under the veil. It is thought that the sotolon is indirectly produced by the yeast when the proportion of alcohol is high. The yeast transforms an amino acid in the wine into a keto acid, which is released with the death of the yeast, falling to the bottom of the cask. A chemical reaction then transforms the keto acid into sotolon, enriching first the bottom, then the middle, and finally the upper layers of the wine.

Now that sotolon is known to be the molecule responsible for the distinctive flavor of yellow wine, research is under way to identify strains of yeast that are capable of producing it in quantity and determine the conditions that favor the emergence of this flavor during aging.

73
Wine Without Dregs

Wines meant for exportation must be stabilized in order to prevent tartrate deposits from forming.

THE COLD OF WINTER CAUSES TARTRATE CRYSTALS to precipitate in bottles of wine in the cellar. These crystals do not harm the quality of a product that holds both symbolic and economic importance for France, but they do hamper its exportation to demanding or poorly informed clients. How can reductions in market value—or, worse still, returns—be avoided? Jean-Louis Escudier, Jean-Louis Baelle, and Bernard Saint-Pierre at the Institut National de la Recherche Agronomique (INRA) station in Pech Rouge and Michel Moutounet at the Institut Supérieur de la Vigne et du Vin in Montpellier have developed a method for balancing wines by electrodialysis.

Tartaric acid is a characteristic constituent of grapes, but the salts associated with it are not very soluble. For this reason potassium bitartrate and calcium tartrate naturally tend to precipitate in wines, forming deposits of what are commonly called tartrates. In the past, producers who wanted to avoid tartrate accumulations kept their bottles at a low temperature for ten days or so, adding to the wine potassium tartrate, which served as a crystal nucleus. The cold triggers crystallization because the limited solubility of tartaric salts diminishes further with the fall in temperature, while the action of polyphenols (a class of molecule that includes many of the coloring agents in red wines) ensures that the wine remains supersaturated with these salts—hence the ineffectiveness of traditional procedures in the case of red wine. Moreover, because the tartrate content of the wine varies, lasting stability is not guaranteed.

The Hunt for Tartrates

Because the cause of the problem is the excessive concentration of tartrate, potassium, and calcium ions in wines, the INRA researchers sought a way to eliminate them. Electrodialysis suggested itself as a promising candidate. Wine was made to flow between two polymer membranes by applying an electrical field perpendicular to the direction of flow: The negative ions were attracted on one side and the positive ions on the other. The researchers reasoned that if membranes were used that selectively let through potassium, calcium ions, and tartrate ions, these ions would be specifically extracted.

The wine was circulated through a stack of parallel anionic and cationic membranes that deionized it while enriching the lateral compartments (through which a solution of fixed composition circulated simultaneously) in tartrate, potassium, and calcium ions. The membranes were made of grafted polysulfones, 20 centimeters (a bit less than 8 inches) on a side and spaced 0.6 millimeter apart, and the electric field applied was 1 volt per cell. The researchers tested a total transfer surface of 4 square meters consisting of sixty stacked cells. In their modeling of the problem they tried to adapt the intensity of the electrical current to the degree of instability peculiar to each wine. As a result, oenologists now have at their disposal a tool suited to the majority of relevant cases.

Why does the new method do a better job of stabilizing the wines than the traditional technique? Because it uses the electrical conductivity of the environment as a measure of the amount of tartrate in the wine. This conductivity depends to a great extent on the concentration of potassium, calcium, and tartrate ions. By regulating the electrical field and the length of time the wine circulates as a function of its conductivity, it becomes possible to extract just the quantity of ions necessary to balance the wine, at a rate of about 1 hectoliter (a little more than 26 gallons) per hour using the pilot device tested. The speed of processing depends on the wine's adhesive properties: Once a day a cleaning agent must be circulated through the system in order to remove the accreted polyphenols and tannins.

Does the new filtering process alter the quality of the wines? This vital question, one that is naturally of great interest to all gourmets, was studied in the course of a long and careful series of trials by oenologists in the Beaujolais, Champagne, and Bordeaux regions. They reported no qualitative difference in flavor between treated and untreated wines. Baccard S.A., in association with Eurodia S.A., which produces the membranes, plans to manufacture and market detartarizing equipment now that the necessary approvals have recently been obtained from Brussels; the first experiments were conducted under private auspices, but preliminary authorization for the expenditure of public funds in support of this research was received in 1996.

In the meantime the researchers continue to study possible uses for the extracted ions and to explore the processing of sweet fortified wines (*vins doux naturels*) and apéritifs made from wine, which are difficult to treat using traditional methods, with the aid of cold.

74
Sulfur and Wine

Sulfur compounds in wine are responsible for defects and virtues alike, depending on the molecule.

IS THE PRESENCE OF SULFUR always a defect in wine? In the 1960s the undue interest of some growers in preserving their wines as long as possible gave sulfur a bad reputation. Sulfur dioxide added in excessive quantities during the fumigation of casks and the sulfiting of harvested grapes causes painful headaches, it is true. But recent biochemical studies show that the use of sulfur is not to be rejected altogether. Chemists at the Faculté d'Œnologie de Bordeaux have discovered that sulfur is capable of both the best and the worst: Although some sulfur molecules are the source of indisputable flaws, others contribute pleasing notes of boxwood, broom, passion fruit, and grapefruit in both white and red wines.

Oenology has long seen only the negative side of sulfur compounds. There is no question that hydrogen sulfide and sulfur dioxide are nauseating. Ironically perhaps, the attempt to eliminate these deleterious effects by improving fermentation and vinification methods led to the discovery of the positive side of sulfur compounds. In 1993, Philippe Darriet and Denis Dubourdieu discovered a molecule in Sauvignon wines having an agreeable odor that belongs to the thiol family (characterized by a group composed of a sulfur atom and an –SH hydrogen atom that is directly attached to a carbon atom). This raised the possibility that other sulfur compounds might contribute to the aroma of Bordeaux wines. Further investigation by Takatoshi Tominaga, Valérie Lavigne-

Cruege, and Patricia Bouchilloux revealed the presence of sulfur compounds in very weak concentrations.

Sulfur and Yeast

Many known sulfur molecules are aromatically very active thiols. Whereas the perception threshold for alcohols (molecules with an –OH hydroxyl group) is on the order of a milligram or a microgram per liter in wine, the threshold for thiols is roughly one-thousandth as much—hence the interest in developing a method of analysis capable of detecting these trace aromas, agreeable and noxious ones alike.

The studies by Tominaga and his colleagues showed that many volatile sulfur compounds responsible for aromatic defects in wines have their origin in the action or metabolism of yeast, which during fermentation transforms the sulfur amino acids of the grape and the sulfur dioxide added as a preservative. This finding suggested a way to correct the known defects of sulfited wines. But whereas it is simple to lower the concentration of hydrogen sulfide, either by reducing the share of sulfur dioxide or by racking the wine in order to aerate it, other molecules such as ethanethiol and methanethiol, as well as a similar molecule known as methionol, are more difficult to control. The chemists observed that the concentration of these compounds depends directly on the turbidity of the grape juice before fermentation. This cloudiness, which brings the juice into contact with the yeast, has to be reduced.

The Misdeeds of Copper

Investigation of the good side of sulfur molecules has yielded its own harvest of results. Certain thiols were known to share the characteristic odor of plants (broom, boxwood), fruits (cassis, grapefruit, passion fruit, guava, papaya), and even foods such as roasted meat and coffee. It happens that the great variety of aromatic nuances found in Sauvignon wines are lost when copper is added. Because copper bonds chemically with thiols and blocks their aromatic action, the Bordeaux chemists suspected that thiols were also responsible for the odors found in Sauvignon.

The first thiol was identified in a red Sauvignon wine, in 1993, and had an odor of boxwood and broom at a concentration of 40 nanograms per liter. The chemists succeeded next in finding it in boxwood and broom as well. They then found other molecules in white wines that are also present in fruits, notably 3-mercaptohexyl acetate, which has a dominant hint of boxwood and recalls the odors of grapefruit zest and passion fruit (in which it was subsequently detected) and another thiol that, depending on its chemical environments, gives off the same fragrances.

And in red wines? Here again the chemists proceeded on the basis of a sensory observation, namely that the complexity and intensity of fruit and meat aromas in young red Merlots and Cabernet Sauvignons decrease when these wines contain a small amount of copper. Several thiols found in Sauvignon wines recently have been identified in young wines made from Merlot, Cabernet Sauvignon, and Cabernet Franc grapes, in which they contribute notes of cooked fruit, cassis, meat, and coffee, at concentrations as low as 0.1 nanograms per liter.

Oenologists are now seeking to use these discoveries to improve wine-making methods. In particular, they know that aging white wines on their lees, where the sulfur molecules of the yeast are found, progressively increases the concentration of certain aromatic thiols.

75

Wine Glasses

The same well-calibrated glass is best for both white and red wines.

DISCIPLES OF THE GRAPE ARE SELDOM without a glass. But which one? Generations of gourmets have debated the optimum form and size of a wine glass. In France, the Institut National des Appellations d'Origine (INAO) has mandated use of a glass recognized by the International Standards Organization (ISO) whose bowl is about twice as high as it is wide. Is it really the best? Because German tasters recommended a rounder glass, Ulrich Fischer, of the Oenology Department at the University of Neustadt, and his colleague Britta Loewe-Stanienda decided to explore the intensity of perceptions of aromas in glasses of various shapes and sizes.

Standards organizations such as the ISO and INAO have stepped in to regulate glassware without having first determined the exact effects of form and shape on perception. It is for reasons of habit, not science, that red wine is served in glasses that are more voluminous than those used for white wine, and mellow wines are served in glasses whose opening is larger than those meant for dry wines.

Calculated Evaporation

Before studying the shape of glasses, the Neustadt researchers sought to determine the physicochemical effects of pouring wine into a glass on the perception of aromas. As a matter of physics we know that the vapor pressure of a molecule dissolved in a liquid depends on the solvent, the solution, and the temperature. We also know that molecules that have passed into the gaseous phase in a glass diffuse into the surrounding air at a rate that depends on the diameter of the rim. On smelling the air in the bowl one inhales these volatile gaseous phase molecules. But many connoisseurs inhale several times in rapid succession. Does another wave of odorant molecules have enough time to pass into the gaseous phase between inhalations? To find out, the researchers used gas chromatography to analyze the air that lies above the wine at different heights and at different times after the wine has been poured into a glass. Some forty compounds were studied at different temperatures.

It was known, too, that the rate of release of the odorant molecules depends on their chemical constitution. Other differences appeared as well: The release rate of esters varied much less, as a function of temperature, than the rate of release of alcohols and volatile phenols. This phenomenon explains why the esters in wines that are consumed when they are cool are concentrated more rapidly than the alcohol, stimulating the perception of fruitiness. Does it also tell us why white wines are drunk at cooler temperatures than red wines? Fruity esters, which often dominate the bouquet of white wines, rapidly present themselves to the nose at low temperatures in sufficient quantities to be perceived, whereas red wines must be served at warmer temperatures for the effects of their volatile phenols to be registered.

Measurements indicated that gourmets are well advised not to quaff their wine at intervals of less than fifteen seconds if they want to have comparable sensations. Their impressions will be strongest if they sniff the air nearest to the rim of the glass.

Tests of Glasses

Having specified the conditions for tasting in detail, the Neustadt researchers tested ten glasses with different maximum diameters, heights, and diameters at the rim. Tasters were instructed to note the intensity of some ten notes (buttery, floral, red fruits, and so on) for the same wine in each glass.

These experiments yielded a great deal of information. With white wines, for example, a narrow bowl brought out the reduced aroma (a faintly vegetal odor suggestive of cabbage) and sulfur character of the wines more than the calix of the ISO glass. Glasses with a large bowl of small or medium height produced weaker sensations than ones with narrow bowls, whatever their height. Although increasing the volume of the bowl did not reduce the intensity of perceptions, glasses with a larger bowl and rim (which are reputed to be particularly well suited for the tasting of red wines) are not superior to the ISO glass, which has a relatively narrow bowl.

Three discoveries in particular upset conventional oenological wisdom. First, the intensity of all perceptions was found to change with the type of glass. Second, increasing the height of the bowl and the ratio between the diameter of the rim and the maximum diameter of the bowl intensified all perceptions. Finally, and contrary to what many manufacturers have claimed, the glass that gave the best results in white wine tastings—the ISO glass—was also the best for red wines. A dogma is slain.

76

Wine and Temperature

Whether one wants to chill champagne or bring wine up to room temperature, it pays to be patient.

EVERYONE KNOWS THAT CHAMPAGNE is best drunk chilled. But how long does the bottle need to be left in the refrigerator? Does it matter whether a child comes along and opens the door of the refrigerator in the meantime? Or say you want to bring wine up from the cellar in order to warm it to room temperature before a meal. How far in advance should you do this? The use of a thermocouple, a device that precisely measures temperatures, gives gourmets useful orders of magnitude.

Let's start with the bottle of champagne. In a simple experiment, the probe of the thermocouple was inserted and the bottle placed in the door of a refrigerator at thermal equilibrium next to bottles that had been there for more than two days and that recorded an internal temperature of 11°C (52°C). The initial temperature of the bottle of champagne was 25°C (77°F). Measurements taken every ten minutes showed a slow cooling: After 30 minutes the temperature of the champagne was still above 20°C (68°F), and it took three hours to reach 15°C (59°F). Only after six hours did it fall to 12°C (54°F), proof that glass is a poor conductor of heat (which propagates in it solely by conduction, whereas in water heat is distributed by convection as well).

Untimely Openings

Suppose you have taken this strong thermal inertia into account, but then a child comes along and opens the door of the refrigerator. Is there a risk that the champagne you have gone to the trouble of chilling all this time will now be warmed up?

This time the temperature in the refrigerator was initially 8°C (46°F). Outside the temperature was 20°C (68°F). When the door is left open for only seven seconds, the temperature rises to 11°C (52°F); after a few minutes it goes back down to 8°C (46°F). When the door is left open for twenty seconds, the temperature rises to 18°C (64°F), then drops rapidly at first and slowly afterward. The air inside the refrigerator heats up very quickly, then, because the cool, dense air inside flows out from the bottom and is replaced by air entering from outside. But this heating up involves only the air itself, not the bottles, whose calorific capacity—and thermal inertia—is substantial. The temperature of the bottles themselves rises hardly at all as a result of a momentary opening of the door, and after a few minutes they cool back down with the rapid drop in the air temperature.

Up from the Cellar

The opposite question, how long it takes to raise the temperature of a wine, has also been neglected. Say you are looking forward to dinner at home with friends and you want to bring up a good bottle from the cellar. Suppose, too, that your guests happen to know that the wine should be drunk at a temperature of 18°C (64°F), but the temperature in your cellar is 12°C (54°F). How far ahead of time should you bring the bottle upstairs?

Naturally one could work out the answer using the laws of physics. But let's conduct a little experiment and measure the actual change in temperature over time. Bring up a bottle of wine whose initial temperature is 9°C (48°F). If the temperature in the dining room is 24°C (75°F), we will find that it takes about half an hour for it to reach a temperature of 12°C (54°F), three-quarters of an hour to reach 14°C (57°F), an hour and a quarter to reach 16°C (61°F), an hour and three-quarters to reach 18°C (64°F), and more than three hours to reach 20°C (68°F).

The long period of time needed to bring a bottle up to room temperature again results from its low thermal conductivity, especially when the difference in temperature between the room and the bottle is small. The answer to our earlier question, then, assuming the same ambient temperature, is that the bottle must be brought up from the cellar more than an hour before dinner is served. This is not a detail to be left to the last minute.

Bottles Versus Glasses

Given these measurements, can we deduce the time needed to bring wine up to room temperature if it is summertime and the temperature inside is 27°C (81°F)? Alas, no; further measurements are needed, together with the equations of thermodynamics. But for our purposes it will suffice to consider certain orders of magnitude. If the temperature of the bottle to begin with is 12°C (54°F), it will take about eighty minutes for it to reach 27°C (81°F). But if room temperature is only 19°C (66°F), it will take more than two and a half hours, starting from the same initial temperature. It also turns out that the upper part of the bottle warms up more quickly than the lower part. The difference may be as much as 4°C (7°F), which produces significant differences in taste between the first and last sips of wine because the higher the temperature, the more quickly its aromas evaporate.

Finally the guests are seated. Once the wine, warmed to 16°C (61°F), has been poured into their glasses, how fast will it heat up when the ambient temperature is 23°C (73°F)? This time, the temperature in a traditional glass increases on average by 0.2°C (0.36°F) per minute, which ought to be enough to stop us from pouring the wine until our friends are ready to hoist their glasses.

77

Champagne and Its Foam

Proteins give champagne its distinctive fizz.

WHEN WE HEAR THE UNMISTAKABLE SOUND of the cork popping off a bottle of champagne, we stop talking and look closely at what happens as it is poured into our glass. If the foam subsides slowly, if the frill of bubbles is delicate and persistent, and if the liquid is effervescent, the wine is considered to be of good quality. Conversely, a rapid fall in the level of the fizz, the absence of a collar, and the presence of large bubbles are taken to be signs of inferior quality, even though the taste of the beverage may otherwise be satisfactory. Champagne makers therefore strive to produce wine that has a delicate and stable foam.

Researchers at Moët et Chandon recently obtained European Union funding to study the physical chemistry of the foam of champagne. In particular they wanted to know which molecules are responsible for its stability and which physical mechanisms account for changes in its structure. At the beginning of the project, evaluation was conducted solely by a panel of judges, a labor-intensive and time-consuming procedure. In short order two devices for obtaining reliable measurements were put into operation. In one, known as a Mosalux, a gas was injected into a fritted glass at the base of a cylinder containing the wine in order to measure dilation and the average duration of the foam. The other, a video system equipped with pattern recognition software, was used to track changes in the structure of the foam in actual champagne glasses.

These devices were first used in connection with filtration studies. Wine growers filter their products to give them clarity and to reduce the concentration of colloids before precipitating the tartaric acid because colloids limit the crystallization of the acid—this despite the fact, as they are well aware, that filtering hurts the quality of their wines, causing them to lose their roundness in the mouth.

How does filtering change the foam of champagne? It is generally thought to be harmful because it eliminates proteins, which are tensioactive molecules. These molecules are composed of hydrophilic parts that dissolve easily in water and hydrophobic parts that avoid water, preferring to be in contact with the air inside the bubbles. Proteins help stabilize egg white foams and the bubbles in champagne: By coating the bubbles they impede the formation of new bubbles and prevent existing bubbles from fusing with their neighbors.

Brewers of beer had observed that filtering did not harm its head, but this is because proteins are much more plentiful in beer than in wine. Research has also shown that proteins with a molecular mass higher than 5000 create a stable head.

In 1990, Alain Maujean and his colleagues at the University of Rheims noted a relationship between the concentration of proteins and foaming capacity in thirty-one wines chosen at random. Yet they were unable to determine the conditions for producing a stable foam. Three researchers at Moët et Chandon, Joël Malvy, Bertrand Robillard, and Bruno Duteurtre, used the Mosalux to study this problem. By subjecting still wines to ultrafiltration, they were able to separate out a part rich in macromolecules (notably proteins) and another part poor in such molecules. Then, by mixing them in different proportions with the base wine, they obtained wines containing various protein concentrations.

Foam Measured

On insufflating these wines with carbon dioxide, the researchers observed the same pattern: The foam initially accumulated and rose in the cylinder of the Mosalux, then subsided a bit, reaching an equilibrium for a time before finally dropping as the gas ceased to be injected. Wines whose protein concentration was reduced by 20–100% exhibited similar foaming behavior, but the higher the content of proteins having a molecular mass greater than 10,000, the greater the measured quantity of foam. Although the bubbling action of a wine during the first seconds was independent of its macromolecular concentration, these macromolecules substantially retarded the foam's decline and the accompanying dilation of its bubbles. As expected, filtering greatly harmed the foam. The disappearance of only a milligram of protein per liter of wine (out of about ten milligrams normally) reduced its ability to foam by half; removal of 2 milligrams of proteins shortened the average duration of the foam by half as well.

These first studies were supplemented by measurements of the foaming power of wines from which colloids and particles had been removed through successive filtrations. The dilation and average duration of the foam diminish by more than half when a wine is filtered by membranes whose pores have a diameter of 0.2 micrometers. With wines that have been aged for more than a year, the consequences of filtering them through pores of the same diameter are still worse, whereas wines filtered through pores 0.45 or 0.65 micrometers in diameter have a more stable foam. This shows that the macromolecules and particles do not act in the same fashion on the bubbles.

How, then, do they act? In the first seconds after injection of the gas, the films separating the bubbles are stabilized by the tensioactive macromolecules. Once the quantity of these macromolecules at the liquid–gas interface exceeds a certain minimum threshold, the foam rapidly coalesces because the bubbles, which contain only carbonic gas, are not in equilibrium with the ambient atmosphere. It may be that other macromolecules in addition to proteins—sugars, for example—act on the bubbles as well. But what is clear—as clear as filtered champagne—is that filtering interferes with foaming and so must be done with care.

78

Champagne in a Flute

Champagne bubbles are more stable in glasses that have been cleaned without a dishwashing detergent.

SPARKLING WINES ARE JUDGED FIRST by the uniformity of their color and by their intensity, clarity, and effervescence. The palate has ample reason to react unfavorably when the eye perceives a bit of tartar or other loose foreign particles. The wines of Champagne nonetheless are measured against a different yardstick. The bubbles that rise up from the bottom of the glass are the most obvious index of quality in the popular mind, together with the accumulation of a fine foam at the top of the glass. The foam should be a few millimeters high, but not more, and the bubbles should be small.

But is the absence of foam a sign of an inferior champagne? Many gourmets are convinced it is. Producers, equally convinced of the quality of their wines, have tended to point the finger at glassmakers, suspecting that the problem arises from the variable surfaces of the glassware into which champagne is poured. Patrick Lehuédé and his colleagues in the research department of the Saint-Gobain group have studied the effect of glasses on the foam of sparkling wines. Their experiments showed that champagne ought not be unfavorably judged by the absence of a foam, which in any case results not from the quality of champagne glasses but from the way in which these glasses are washed and stored.

In theory the formation of the bubbles is simple. Champagne is not in stable equilibrium when it is first opened and comes into contact with the air outside the bottle. The gas escapes from the liquid, forming bubbles either on flaws present in the body of the glass—scratches, for example—or, more commonly, on the surface of the glass. The size of the bubbles depends on the energy of the surface coated by the liquid, which is to say the ease with which a surface accepts contact with other materials, whether liquid or gaseous. Once formed the bubbles grow in size because the pressure of the gas in the liquid is greater than the pressure inside the bubbles, with the result that the gas molecules diffuse toward the bubbles. Eventually the Archimedean thrust exerted on the bubbles exceeds their force of adhesion, and the bubbles detach from the sides of the glass and slowly begin to rise.

This rough description can be refined. The adhesive energy of a bubble clinging to the glass is proportional to the surface area of contact between the bubble and the glass, which in turn depends on the surface energy of the glass. Surface energy is conventionally measured with respect to the angle of contact of a water droplet with the glass. When this energy is great the surface is well moistened by the liquid, the contact area is small, and the bubble is quasispherical. The bubble then becomes detached when its diameter reaches a few tenths of a millimeter. This is what happens with most ordinary sodium–calcium glasses when they are clean. By contrast, when the surface energy is low, which is to say when the liquid does a poor job of moistening the solid, the bubble detaches itself only once it has become large (more than a millimeter in the case of certain plastic materials). In other words, gourmets who refuse to drink from plastic flutes are right: Champagne bubbles are smaller in a glass flute.

Theory Tested by Bubbles

The Saint-Gobain physicists first studied the formation of bubbles using various microscopic methods. Contrary to a common belief, bubbles do not form on flaws in the glass itself, for scratches are rare. Microscopy demonstrated that they appear mainly on limestone and tartar deposits and on cellulose fibers left on the surface of the glass by towels used in drying. The proof? Glasses free from any deposit, prepared in a clean room and immediately filled with champagne, do not allow a single bubble to form.

Nonetheless, Lehuédé and his colleagues showed that high-energy surfaces are rapidly contaminated (within a few hours in the case of an ordinary glass) by organic molecules that are present in the air (these abound in kitchens, as you can see by running your finger along the ceiling above the stove). The inherent advantage of glass and crystal over plastic therefore is reduced when flutes have gotten dirty from sitting for a long time in an inappropriate environment, such as on wooden shelves, which release organic essences (as you can readily tell by smelling the wood).

The bubbles that detach themselves from the surface of traditional champagne glasses once they have become sufficiently large rise and feed the collar of foam, whose height depends not only on the number and dimensions of the bubbles but also on their stability. Certain compounds are well known for their power to modify this stability, such as the antifoaming agents typically found in red lipstick (so that frothy substances do not adhere to the lips), which explains why people wearing lipstick have less foam in their champagne glasses after the first sip. The effect is spectacular, as you can see for yourself by touching the foam in a glass of champagne with the tip of a lipstick.

Researchers at Moët et Chandon discovered that detergents frequently used in dishwashers have the same effect. Adsorbed on the surface of a flute, these compounds are dissolved when it is filled with champagne. Although they do not disturb the wine's effervescence, they do affect the stability of the foam, causing the bubbles to burst when they reach the surface. Dishwashing liquid, though also harmful, is less of a problem because rinsing eliminates most of it.

79
Demi Versus Magnum

Champagne ages more quickly in small bottles.

THE MAGNUM OF CHAMPAGNE is a prince among princes: Connoisseurs ascribe virtues to it that they do not detect in regular or half-bottles of the same wine. Are they correct? And are they correct in supposing that champagne should be stored lying down because the cork remains moist and so better preserves its hermetic properties? Experiments performed recently by Michel Valade, Isabelle Tribaut-Sohier, and Félix Bocquet at the Centre Interprofessionnel des Vins de Champagne (CIVC) in Épernay led to a new understanding of the role of corks in the aging of sparkling wines.

Champagne is a wine that foams, by contrast with still wines, in which the presence of bubbles is considered almost a flaw. Why does champagne foam? Because the yeast put into the wine consumes its sugar, releasing carbon dioxide in the space enclosed by the bottle. This gas is then dissolved in the liquid, causing it to bubble when the cork is popped. Along with the bottle itself, the cork is the key to preserving the qualities of a good champagne because it ensures that the wine retains its gas.

Cork Scrutinized

For a long time it was supposed that corks are perfectly hermetic. After all, the wine does stay in the bottle. The wine, yes, but not the gas: As oenologists well know, bottles of champagne lose their pressure over time. This phenomenon led them to examine the behavior of corks more closely. In champagne making, before the familiar mushroom-shaped cork is inserted, producers use crown caps equipped with a temporary seal that can be removed to add sugar. This seal comes in either cork or plastic (a synthetic polymer derivative). The CIVC team studied the two types of material and found that cork seals were not uniformly impermeable, which explains the variations that are observed from bottle to bottle.

Synthetic seals displayed a consistently higher degree of impermeability, but this fact alone does not establish that they are the best device for blocking the escape of gas. However, tasting juries have unanimously found that wines with plastic stoppers change less quickly and have less of the cooked fruit taste that is often associated with oxidation. This raises a series of questions. Why does oxidation occur in the first place? Does oxygen diffuse through the stopper? Does the surface area of the stopper in relation to the volume of liquid explain the differences tasters detect between half-bottles, bottles, and magnums?

Tests of Aging

An impression—even a highly educated one—is not the same thing as a controlled experiment. To determine whether the size of the bottle really makes a difference, the Épernay oenologists compared samples of the same champagne, drawn and bottled under identical conditions, and plugged in the classic fashion using crown caps with cork seals. Then, after a year of aging, they submitted the wines to the judgment of a panel of tasters. Consistent with earlier results, the wine in the magnums seemed younger than that in the regular (75-centiliter) bottles, and the wine in these bottles seemed less developed than that in the half-bottles.

Could these differences have resulted from a chemical reaction between the wine and the oxygen present in the small volume of gas trapped when the bottle was capped? No, for measurements showed that the oxygen trapped in this volume and subsequently dissolved in the wine is completely consumed by the yeast, leaving only a residue of nitrogen and carbon dioxide.

Are the changes in the wine caused by a gaseous exchange with the atmosphere outside the bottle despite its being plugged with a cap? Measurements showed that the quantity of oxygen that enters through the stopper during aging is proportionally greater when the capacity of the bottle is small: Although the quantity of oxygen is roughly identical for the three types of bottles, the smaller the bottle, the smaller the volume of liquid that reacts with this oxygen. This is the advantage of the magnum: The larger the bottle, the less effect oxidation has on the wine. But even if it is clear why carbon dioxide manages to escape the bottle, why does oxygen get in? Because its partial pressure outside the bottle (about a fifth of an atmosphere) is greater than the partial pressure inside the champagne (equal to zero).

The Position of the Bottle

As for the position in which a bottle is kept after it has been purchased, this has no influence at all—except on the cork, whose mechanical properties are better preserved when the bottle is standing rather than lying down. More precisely, the force needed to extract the cork is greater in this case, with the cork reassuming its mushroom shape on being removed from the bottle. When the wine is in contact with the cork, by contrast, it gradually penetrates the cork and alters its mechanical properties.

Finally, the CIVC team explored the influence of the moon on the increase in sugar in grapes before harvest. Preliminary findings confirmed what had long been suspected: It has no effect whatever.

80

The *Terroirs* of Whiskey

Statistical analysis provides guidance in scotch tasting.

SCOTLAND, BLESSED LAND OF WHISKEY! There it is called scotch, and there are several kinds. Single malts are made from fermented barley at a single distillery; blended whiskeys are mixtures of different kinds of whiskey that may come from different parts of the country. Obviously the single malts are preferred by connoisseurs, who scrutinize them with regard to five crucial qualities: nose, color, body, mouth, and finish.

Does the provenance of a single malt determine its organoleptic qualities? If it does, can types of scotch be associated with particular environments (or *terroirs*)? To find out, François-Joseph Lapointe and Pierre Legendre at the University of Montreal analyzed data collected by taster Michael Jackson, who has described, tasted, and judged the roughly 330 single malt whiskeys produced by 109 of the most prestigious distilleries in Scotland.

Characteristics of Single Malts

Statistical analysis of single malts proceeds from the fact that their distinctive characteristics assume several forms. Thus the nose may be aromatic, peaty, light, sweet, fresh, dry, fruity, grassy, salty, sherry-flavored, spicy, or rich. The body may be smooth, medium, full, round, honeyed, light, firm, fat, and so on. What is the best way to divide the population of single malts into groups of analogous individuals in order to determine whether they come from the same *terroir*? The University of Montreal statisticians simplified the problem by considering only one scotch per distillery, for a total of 109 samples.

To quantify the various sensory impressions, or characteristics, produced by these scotches, they assigned to each one a value of 1 if it was present in a single malt and 0 if not. They then displayed the statistical data in the form of a table by arranging the individual scotches in rows and the characteristics in columns.

Next they calculated the numerical distance between pairs of single malts by dividing the number of characteristics common to the two whiskeys by the total number of characteristics identified in either one of the two; the distance between them therefore is equal to 1 minus this parameter. The set of these distances was then used to form a new matrix, where the value of a given compartment represents the distance between two whiskeys linked with one another by row and column. The smaller this distance, the closer the whiskeys to each other.

In order to classify individual whiskeys, Lapointe and Legendre partitioned the population by aggregating the closest individuals in pairs. This aggregate is then considered to be a new individual that replaces the two aggregated individuals, whose characteristics are averaged to arrive at the value of the new individual's characteristics. Proceeding in this fashion until all individuals have been related to one another as members of a single family, one winds up with a dendrogram (or tree diagram). Different partitions can be obtained by making a "cut" at a given distance from the root: The nearer to the root the cut is made, the smaller the number of classes.

In Search of Class

The dendrogram obtained for single malts divides into two branches. On one side are golden whiskeys having a dry, smoky body. On the other are amber whiskeys that have a light body, smooth in the mouth with a fruit finish. The first group, consisting of 69 whiskeys, was then subdivided into amber scotches, which are full-bodied, fruity, fat, and spicy, and golden scotches, which have a smooth and light body with a grassy finish. The farther away one goes from the root, the smaller the groups.

Let's come back to the geographic component. Scotland is divided into three scotch-producing regions: the Highlands, the Lowlands, and the Isle of Islay.

These three regions are divided into thirteen districts. The researchers constructed a new table in which the 109 scotches were again arranged by row and column. The compartment at the intersection of a row and column contains a 0 if the corresponding scotches are from the same district, otherwise a 1.

Comparing this matrix with the one that generated the dendrogram, one observes that the division into regions of production corresponds to the tree diagram division, whether six or twelve groups are considered. These groups correspond to *terroirs*. What is more, this distribution confirms that the water, soil, microclimate, temperature, and overall environment are indeed the determining factors of the characteristics of single malt whiskeys, as upholders of the *terroir* theory maintain. The secrets and traditions of individual distillers account for only small differences by comparison.

Finally, the Canadian researchers sought to relate the five types of characteristic to one another. It was clear that nose, color, body, and mouth are not independent characteristics: The color of a scotch is related to its nose and its body, and the nose is related to the mouth and, to a lesser degree, the body. The finish, on the other hand, which is to say the impression that is left in the mouth when one has finished drinking, depends on neither the nose, the mouth, the body, nor the color. The familiar method of tasting that consists of spitting out the beverage after its color, nose, body, and mouth have been judged therefore is open to criticism because it neglects a fundamental parameter.

81

Cartagènes

Where the aroma overtakes the alcohol.

TRADITIONALLY MADE BY ADDING ALCOHOL to fresh, unfermented grape juice, the *cartagène* of Languedoc has never enjoyed the reputation of the Pineau des Charentes or other *mistelles,* as such apéritifs are known in France. Nonetheless, its producers sought to obtain a protected designation of origin, which meant that it had to be more precisely characterized and its method of production codified. At the request of the manufacturers, Jean-Claude Boulet and his colleagues at the Institut National de la Recherche Agronomique (INRA) Pech Rouge-Narbonne station studied the importance of the grapes used and the conditions of maturation.

In *mistelles,* the fermentation of grape juice is blocked by the addition of brandy (equal to one quarter of the volume, hence the name *cartagène*). The strong concentration of alcohol (at least 16%) prevents the microorganisms that normally bring about the alcoholic fermentation of grape juice into wine from developing.

Yet the taste of alcohol in young *mistelles* is too strong, almost like pure brandy. To obtain a more pleasing result, producers favor a slow and limited oxidation of the polyphenols (molecules containing benzene groups in which several carbon atoms are bonded with hydroxyl groups, themselves composed of an oxygen atom linked to a hydrogen atom), tannins, and other molecules extracted from the grape during the short period of maceration (or steeping) that follows pressing. This oxidation process resembles the one responsible for the softening of tannins in wines and results from reactions with the small quantity of oxygen that is inevitably present in fermentation vats.

To study the oxidation of *cartagènes,* the INRA oenologists made sample batches of the apéritif in a cellar laboratory using three grapes: Syrah, Grenache, and Cinsaut. There were ten carefully controlled phases of production: removing the stalks from the grapes; macerating the skins in the grape juice for four hours at room temperature; pressing; fining; leaving the wine to settle for two to three days at a temperature of 5°C (41°F); racking, or drawing off clear wine from the sediment; adding high-quality white brandy to the wine; storing the mixture for a month; a further round of clarification; and, finally, aging the wine in a stainless steel vat or an oak barrel that used to contain cognac. The INRA team avoided sulfiting (the addition of sulfur dioxide, often used in wines to kill microorganisms or to take advantage of its antioxidant properties) and included a maceration step to extract the most aromatic molecules and polyphenols from the skins. Finally, the researchers compared other *cartagènes* produced by Languedoc wine growers from (white) Bourboulenc and (red) Alicante grapes with the laboratory batches.

Useful Aging

The *cartagènes* produced in the lab were subjected to physical and chemical analysis and sampled by a tasting jury after different maturation periods. The oxidation of certain initial molecules was obvious: The color of the liquid—white if made from white grapes, red if from red grapes—gradually became uniform. Chemical analysis showed that the polyphenols and tannins were a bit more abundant in casks than in vats. Despite the use of old barrels that had previously contained brandy, exchanges still occurred between the *cartagène* and the wood, yielding various polyphenols.

During the testing phase, the tasters did not notice these various polyphenol concentrations. By contrast, they were very sensitive to the differences between young and aged batches. In *cartagènes* held for only six months, the one made from Syrah was appreciated for its fine red color and its fruity, nonoxidative character; the ones made from Grenache and Cinsaut were less colored and less aromatic, displaying no perceptible differences associated with the mode of maturation. In all three cases the alcohol was very much in evidence, indeed harsh. Yet after fifteen months no difference could be detected between batches made from Cinsaut that had been aged in casks and those aged in vats; the difference was small in the case of Grenache but larger in the case of Syrah. All three types were sufficiently oxidized.

Were the judges' findings reliable? The Narbonne biochemists confirmed first the general coherence of their responses, but they observed that the tasters fell into three groups: members of the INRA station, none of them experts on *cartagènes;* producers' representatives; and people who gave atypical responses. Discarding the "bad" tasters of the last group, they arranged for another tasting at which the *cartagènes* served were younger. Once again the results of the initial tasting were confirmed, but this time no difference was perceived between *cartagènes* of the same age, whether they were aged in vats or in casks.

On this round of tasting the Syrah seemed to yield a fruity and agreeable drink more rapidly than the other groups; the *cartagènes* made from Cinsaut and Grenache acquired a distinctive style only through limited oxidation of polyphenols and other oxidizable molecules, but for this reason they surpassed the Syrah *cartagènes* in aromatic power. In all three cases aging for at least a year was found to be indispensable. A longer aging period would increase production costs, but whether it would further improve quality remains to be demonstrated.

82

Tea

The chemistry of clear plaques on the surface of tea.

FOODS AND BEVERAGES contain so many amphiphilic molecules—one part of which are water soluble and another part insoluble—that foams are common in the kitchen: stiffly beaten egg whites, champagne, beer, cream, and so on. In cooking one sometimes tries to minimize contact with antifoaming agents, taking care not to spill any drops of egg yolk into whites that are about to be whisked, for example, because the fatty molecules in the yolk bond with the hydrophobic part of the proteins in the white, removing them from contact with the air and thus stabilizing the water–air interface. Similarly, people who like the fizz of champagne are careful not to wear certain kinds of lipstick that contain antifoaming molecules. Under other circumstances, however, foam is something to be avoided. When making apricot jam, for example, one can cause the froth created by cooking to subside by daubing the surface with melted butter.

Films Disturb Connoisseurs

A pair of chemists at Imperial College, London, both great lovers of tea, set out to investigate the thin clear plaques that appeared on the surface of their teacups. Michael Spiro and Deogratius Jaganyl initially supposed the plaques to be the residue of a foaming phenomenon that had been arrested by the hardness of the water, but they were surprised to find something entirely different, which they described in an article in *Nature*.

The films that one observes in teapots and teacups are irregularly shaped plaques that to the naked eye look like surface stains. Spiro and Jaganyl were able to study them quantitatively once they had devised a method for collecting them from the surface of large containers that they had infused with tea. Examination of these films with a scanning microscope revealed the presence of small, clear particles on their surface that turned out to be calcium carbonate. Microchemical analysis confirmed that all of the calcium, as well as traces of magnesium, manganese, and other metals coming partly from the municipal water supply and partly from the tea, could be eliminated by treatment with chlorhydric acid. The remaining particles were organic in nature and insoluble in all the solvents tested but soluble in concentrated bases. Mass spectrometry indicated that this residue was composed of molecules of different mass, in the neighborhood of 1000 daltons.

Spiro and Jaganyl also studied the rate of formation of these films by infusing black Typhoo tea in water heated to exactly 80°C (176°F) for definite periods of time. After the infusion phase they removed the tea bags, skimmed any froth from the surface, and then measured the formation of films over time. They observed that films formed for several hours, their quantity increasing proportionally with the number of minutes elapsed during the first hour; after four hours, the mass of the thick film that had formed was proportional to the surface of the container. The rate at which the film formed depended on the atmosphere above the tea: The mass of film that appears in atmospheres of pure oxygen is greater than that which develops in ordinary air or under a nitrogen atmosphere. It was clear that the film was produced by the oxidation of the tea's soluble compounds, certain polyphenols among them (the bitterness of tea results from such molecules).

The British chemists did not observe any film at the surface of tea infused in distilled water or in distilled water to which calcium chloride had been added. A film appeared only if the water contained both calcium (or magnesium) and bicarbonate ions. Nor did any film form when the calcium ions were sequestered by a chelating compound such as ethylenediaminetetraacetate or when the tea was acidified. No film appears in tea containing lemon, for example, because its acidity is greater than that of unadulterated tea, its calcium ions being sequestered by citrate ions. Similarly, very strong teas have little film, for the abundance of polyphenols increases their acidity. By contrast, the addition of milk greatly increases the amount of film, as does raising the temperature of the infusion.

The complex organic compounds found in such films, which result from the oxidation of soluble molecules in the presence of calcium ions and bicarbonate ions, seem to form by a process analogous to the enzymatic oxidation used by tea producers to transform green tea into black tea. Further research will be needed to confirm this hypothesis.

4 A Cuisine for Tomorrow

PART FOUR

83

Cooking in a Vacuum

New devices can improve traditional culinary techniques.

COOKS FILTER STOCKS TODAY just as they did in the Middle Ages: They put the bones and vegetable matter in a conical strainer known as a chinois (or China cap) and press it with a pestle or a ladle to squeeze out as much liquid as possible. Naturally the effectiveness of this procedure is limited by the size of the mesh of the chinois. A fine cloth liner helps, but it has to be cleaned after every filtering. Can't we devise a more modern and efficient method? Looking to chemical laboratories for inspiration would be a useful first step on the road to culinary innovation.

Filtration in the Lab

The chief problem encountered in making a good stock is primarily a question of filtration: What is the best way to make a cloudy liquid clear? Traditionally clarification has been achieved by stirring a few egg whites into the cold stock and then heating the mixture over a low flame so that, when they coagulate, the whites trap the solid particles suspended in the liquid. Straining the mixture through a chinois lined with linen completes the process.

This procedure is unsatisfactory because it robs the liquid of a part of its flavor. Some chefs therefore add vegetables and fresh meat, cut into small pieces, along with the egg whites, to restore the flavor lost through clarification—a costly business. Imagine going to the trouble of cooking a stock for several hours and then having to re-enrich it because cooking has impoverished it.

Chemists, for whom filtration is a daily activity, solved this problem long ago with the aid of various devices adapted to specialized purposes. Indeed, the catalogue of one leading supplier of laboratory equipment today devotes more than forty pages to such devices. One of the most commonly used models has a funnel equipped with a fritted glass plate (which, unlike paper filters, does not tear) containing pores of uniform size. The matter to be filtered is deposited in the funnel, and the funnel is then placed on top of a conical vial in which a vacuum has been created by means of a waterjet pump, an inexpensive device that attaches directly to a faucet.

Antoine Westermann, the chef at Buerehiesel in Strasbourg, and I tested this apparatus with a tomato consommé that he wanted to be perfectly clear. The original recipe called for cooking the tomatoes in water to which egg whites had been added. After a half hour of slow cooking, straining the mixture through a cloth-lined chinois yielded a golden liquid. The laboratory device, on the other hand, yielded both a clearer liquid and a more pronounced taste. What prevents the makers of electric household appliances from producing this piece of equipment on a large scale? They would have only to increase the filtration capacity of the laboratory device and replace the glass with metal that will stand up to rugged use in commercial and home kitchens.

Other culinary uses for waterjet pumps can readily be imagined as well. Nicholas Kurti thought to use one to produce a new kind of meringue. Classically one makes meringues by adding sugar to egg whites that have been beaten until they are stiff and then cooking this mixture over very low heat. The coagulation of the proteins in the egg whites conserves the alveolar structure of the "floating islands," while the water slowly evaporates, leaving a vitreous sugar residue. Kurti had the idea of substituting vacuum storage for heating. In this case the water evaporates while the dilation of the air initially present in the bubbles causes the meringues to greatly expand. The final result is light and airy—like "wind crystals."

From Meringue to Soufflé

Although it has the virtue of showing how vacuum techniques can be used in cooking, this procedure is not altogether satisfactory, for the new meringues are *too* light and airy—there's nothing to bite into. On the other hand, we can increase expansion and have something left to eat afterward if we choose a preparation in which the walls between the bubbles are thicker than in meringues. Soufflés and cream puff pastries come to mind as attractive candidates.

If one combines the ingredients needed to make a soufflé—flour, butter, milk, and eggs—in a vacuum bell jar the mixture swells up when the air is pumped out, for the air bubbles in the preparation expand, but the soufflé collapses when it is put back in atmospheric pressure. To prevent this from happening one needs to cook the soufflé in its expanded state—for example, with the aid of an electric heating element wrapped around the ramekin—in the vacuum bell jar. Heating it in this way causes the proteins to coagulate and the bubbles to swell up with water vapor, so that the soufflé preserves its structure when it is put back in atmospheric pressure.

No doubt many other uses for vacuums will be found once cooks decide to exchange medieval for modern equipment.

84

Aromas or Reactions?

Two ways of imparting flavor to food.

"THINGS OUGHT TO TASTE LIKE WHAT THEY ARE," the gastronome Curnonsky used to say. His aphorism has been adopted as a slogan by those who seek to promote authenticity in cooking, but does it really make sense? Isn't the role of the cook to transform foods with the purpose of recreating traditional dishes and inventing new ones?

If the true aim of cooking is to produce specific flavors, the question arises how to incorporate them in various dishes. There are two ways: by adding flavors or by organizing chemical reactions in such a way that flavors are formed in the foods themselves. One technique that has been widely used by the food processing industry involves both natural extracts and synthetic molecular solutions. The use of these so-called aromatic preparations in cooking is straightforward (one simply adds a few drops to the food), but devising them takes the same kind of technical artistry possessed by the "noses" of the perfume industry, laboratory chemists who concoct novel solutions of various odorant molecules in order to approximate or reconstitute familiar scents such as strawberry, ginger, and rosemary.

Cooks are understandably reluctant to allow themselves to be supplanted by such technicians, all the more because the use of natural ingredients (real thyme and real rosemary in a ratatouille, for example) often gives a richer, and certainly more varied aromatic result than artificial thyme or rosemary flavoring (which usually do not contain as many aromatic molecules as natural ingredients).

Must we therefore dismiss such aromatic engineering altogether? This would mean foregoing the opportunity to enlarge the palette of flavors. Why not reinforce the green note of olive oil with hexanal, or add 1-octen-3-ol to a meat dish in order to give it an aroma of mushroom or mossy undergrowth (although here one needs to be careful about proportions because in excessive concentrations the same molecule smells a bit moldy)? Why not use beta-ionone to give desserts the surprising violet aroma that flowers have such a hard time releasing?

Cooks would be also able to create taste, rather than flavor, by using monosodium glutamate and other molecules that impart the taste called umami, which is naturally contributed by onions and tomatoes. They would be able to use licorice or glycyrrhizic acid, which communicate specific tastes that are neither salt, sugar, sour, nor bitter—nor umami.

Advanced Uses of Fire

Need we worry that in this case culinary progress would be limited to chemical aromatization, which is only a modern version of adding fines herbes and spices to foods? Certainly not: Cooks well know that cooking transforms the taste of their creations. Fire is their inseparable ally, and chemistry can help them make the most of it.

For example, one can easily change the flavor of caramel by varying the type of sugar. Caramels can be made from glucose, fructose, or, more generally, from sugars other than sucrose (ordinary cane sugar). An experiment that anyone can perform will show that these caramels may already be present in foods. It is based on an apparently paradoxical observation made by cooks who, in the course of making a béarnaise sauce, for example, reduce a combination of chopped shallots and white wine until the liquid is completely evaporated: Certain white wines leave no residue in the pan. Why? Because they lack glucose, glycerol, and many other things. The practical lesson is this: If your wine is insufficiently rich in such aromatic molecules, add some glucose to it before reducing *à sec* and you will obtain a glucose caramel that improves the flavor of the sauce.

The Paradox of Reductions

There is something puzzling about even a reduction that has been fortified in this way, however. Why should one want to evaporate most of the aromatic molecules that are present in wines? The same question arises in the case of stocks, which are made by concentrating beef broths through heating. If this concentration has the effect of eliminating the aromatic molecules, why are stocks nonetheless fragrant and flavorful?

Anthony Blake and François Benzi at the Firmenich Group in Geneva used chromatography to compare a stock that had been reduced by three-quarters and then restored to its initial volume by the addition of water with the same stock that had not been topped off. Although the concentration of certain aromatic compounds was reduced by boiling, other compounds were created by heating-induced reactions between the components of the stock. It remains to identify these reactions in order to improve the making of stocks, if possible.

On Chemistry in Cooking

The possibilities of chemistry are unlimited. Our kitchen shelves hold a great many nearly pure ingredients: sodium chloride, sucrose, triglycerides (in oil), ethanol, acetic acid, and so on. And the shelves of our libraries contain a great many chemical treatises that perfectly describe the reactions of these molecules. The challenge facing cooks and chemists today is to apply this knowledge in order to create new flavors.

Take Maillard reactions between amino acids and certain sugars, which produce the tasty brown compounds in the crust of roast beef and bread as well as the aromas of coffee and chocolate. Chemists know that these reactions differ according to the acidity of the reactional environment. Why not soak chicken breasts in vinegar or bicarbonate of soda before putting them under the broiler? To obtain the right degree of acidity once the reaction has occurred, the vinegar could be neutralized with bicarbonate of soda or vice versa.

85

Butter: A False Solid

How to make it spreadable.

BUTTER IS A STRANGE SOLID: When one takes it out of the refrigerator it is sometimes necessary to wait as long as fifteen minutes before it can be spread easily. Would it be possible to make a butter that is spreadable immediately after being removed from the refrigerator?

The question has been around since 1988, when legislation in France granted a butter appellation to products that, like butter, consist of droplets of water in milk fats, on the condition that such products have been separated by physical methods. Thus one could imagine selling butters having various properties, prepared by mixing together various ingredients that were first isolated from butter.

How would one go about separating and recombining these ingredients? In 1992 the research department of the dairy manufacturer Arilait hired several laboratories to analyze milk fats and to identify the physical principles governing spreadable butters. The analysis was complicated by the fact that the molecules that make up milk fats are various and polymorphous; that is, each type of molecule crystallizes in several ways, depending on the sort of processing it has undergone, and the crystals assume their equilibrium form only after a long resting period.

In milk the fatty matter assumes the form of droplets dispersed in water. Each droplet is a few micrometers in diameter and coated with casein micelles, each micelle being a collection of several proteins cemented together by calcium phosphate. Aromatic molecules are dissolved in the fatty matter of the droplets, and other molecules (vitamins, sugars such as lactose, mineral salts, proteins) are dissolved in the water of the milk.

A team sponsored by Arilait first studied this fatty matter, whose composition changes even with the seasons. But there is unity as well as diversity: The fatty matter in milk is made up of triglyceride molecules, composed of a glycerol molecule to which three fatty acid molecules are bound (although, of course, the glycerol and fatty acid molecules lose their identities, as when oxygen and hydrogen molecules combine to form water). On the other hand, milk contains more than 500 such fatty acids, and because each acid can bind with any carbon atom in glycerol, the number of possible triglycerides exceeds several thousand.

Strange Fusion of Butter

One of the chief consequences of this diversity is the strange behavior of butter at the point of fusion. Unlike a pure body such as water, which melts at a fixed temperature (0°C [32°F]), the fusion of butter begins at –50°C (–58°F) and ends at about 40°C (104°F). The various triglycerides in milk melt in three principal stages involving homogeneous chemical families: From –50°C (–58°F) to 10°C (50°F) one observes the fusion of molecules whose fatty acids are short and composed of double chemical bonds among carbon atoms; between 10°C (50°F) and 20°C (68°F) one sees the fusion of molecules containing a single double bond or a short chain; finally, between 20°C (68°F) and 40°C (104°F) one sees the fusion of molecules that contain three fatty acids and have only single chemical bonds between carbon atoms.

Instead, physical chemists Frédéric Lavigne, Michel Ollivon, and their colleagues in the faculty of pharmacy at Chatenay-Malabry used melted butter to study the opposite of fusion: crystallization. To separate the various parts they therefore performed a split crystallization, slowly cooling the liquid and isolating crystals of the same molecular type that appear at the same temperature.

New Butters

Having thus isolated these similar parts, Lavigne, Ollivon, and their colleagues next looked for a way to form mixtures that would be spreadable straight out of the refrigerator. They hit on the idea of mixing high–fusion temperature triglycerides, which remain solid at room temperature, with a suitable proportion of low–fusion temperature triglycerides, which are liquid at room temperature.

In this way one obtains an apparently solid body that, like traditional butter, contains a proportion of molecules in liquid form (even in milk the fatty droplets are partially solid, the proportion of solid matter reaching 70% at 4°C [39°F] but only 10% at 30°C [86°F]). Enrichment by low–fusion temperature molecules makes the mixture easier to spread. The parts that fuse at high temperatures are used to make pastry (still under the name *butter* because the law permits it), particularly puff pastry.

Why, then, does a solid that contains liquid appear to be solid? Because of the crystals that increase with cooling and interlock with one another: Scraped with a knife, butter seems to soften, not because it is heated but because the crystals are separated.

To have an idea how these discoveries can be used in cooking, try testing split crystallization yourself. Melt the butter and skim off the solids as they form, just as the physical chemists did. You will then be able to manufacture your own butters by mixing proportions of solids and liquids and in this way obtain the specific texture appropriate to a particular dish.

86

Liver Mousse

Its aromatic qualities depend on its texture.

THE FOOD INDUSTRY IS FOREVER LOOKING for ways to make its products lighter by reducing lipid content and increasing water or air content, without the taste suffering as a result. Even liver mousse, renowned for its sturdy lipidic constitution, has not been spared. Michel Laroche and his colleagues in the Laboratoire d'Étude des Interactions des Molécules Alimentaires at the Institut National de la Recherche Agronomique (INRA) station in Nantes studied liver mousses to which a large amount of starch had been added, in order to determine by how much the fat content of such products can be reduced without affecting flavor. Physicochemical measurements and sensory analyses showed that the attractiveness of lighter versions depends principally on their meltability, that the sensation of fattiness does not depend on the quantity of lipids substituted for by starch, and that the perception of flavor depends on their consistency.

Making a low-fat liver mousse is particularly tricky because one wants it to be easily spread on toast, a property that in classic mousses seems to result from their high lipidic concentration (as much as 50%). Earlier, in 1985, two other researchers from the Nantes station, René Goutefongea and Jean-Paul Semur, showed that this property could be preserved by adding hydrocolloids, which is to say dispersions of long molecules in water. The recent experiments tested the partial or complete substitution of lipids by starch paste obtained from fava beans, the seeds of a leguminous plant (*Vicia faba*) cultivated in France, which preliminary studies suggested might be a suitable candidate.

In a liver mousse prepared in the classic fashion, by grinding up pork livers with egg whites, lactoserum, gelatin, sodium nitrite, salt, pepper, onions, shallots, and cognac, the Nantes researchers replaced the various quantities of lipids with fava bean starch paste (15% starch in water) for a first batch of samples and with a fixed quantity of paste having variable concentrations of starch for a second batch (so that the result would be 50% lighter but with different starch concentrations). These preparations were compared with liver mousse prepared by traditional methods that was purchased at a local grocery store.

Various mechanical measurements were made to characterize these different mousses, which were then tasted by ten trained judges. In a room dimly illuminated by red light, the tasters were instructed to evaluate four sensory elements: meltability (defined as the ease with which a sample melts between the tongue and palate), fattiness, granular texture (defined as the perception of particulate matter), and intensity of flavor. Finally, they gave an ordered ranking of these elements with respect to their relative contribution to the overall perception of quality.

First, meltability was found to increase with the quantity of starch in the first batch of substituted products but to diminish in the second. The perception of granular texture was not changed by the amount of starch unless it was very large. And the sensation of fattiness, which was independent of the actual quantity of fat for the batch with variable lipid content, diminished by contrast when the starch content was increased in the second batch. An analogous variation was observed for the perception of flavor, which increased for the first batch but diminished with the quantity of starch in the second batch.

The four sensory indices turned out to be strongly correlated, but the relationship between meltability and the perception of flavor is the most interesting: The increase in the quantity of starch in the second batch was associated with diminished meltability and a diminished perception of flavor. Should it be concluded, then, that the aromatic compounds were adsorbed by the starch? Or that they were adsorbed by the water? No; both of these conclusions are invalidated by the results obtained for the first batch, where an increase in the quantity of starch went hand in hand with an increase in the water content of the mousse and a notable increase in meltability.

Because the composition of the samples does not explain the relationship between meltability and aromatic quality, it seems either that the increase in meltability, which is associated with a more even distribution of the mousse in the mouth, improved the perception of flavor, or that the less smooth the texture, the more trouble the tasters had perceiving the other elements.

Meltability was well correlated with high marks for the other sensory indices, in order of importance for the overall impression of quality: meltability, fattiness, flavor, and granular texture. These four parameters were in turn strongly correlated with the measurements of hydration and mechanical resistance. What is the optimal proportion of starch? Using a 15% starch solution, one can replace two-thirds of the lipids without diminishing the overall quality of the mousse; at higher concentrations, however, the mousse becomes too soft.

These studies lead us to conclude three things. First, meltability is the chief factor in determining the overall quality of a liver mousse. Second, the sensation of fattiness is independent of the quantity of lipids that are replaced. Third, the perception of flavor depends on texture. This final result calls to mind the finding of Patrick Étiévant and his colleagues at the INRA station in Dijon, in 1990, that the addition of pectins to strawberry jam firmed up the texture while also reducing its aromatic qualities.

87

In Praise of Fats

Whatever else may be said about them, fats are to be welcomed in cooking.

FATTY FOODS ARE ACCUSED OF BLOCKING our arteries and making us fat. Unsurprisingly, perhaps, one hears calls nowadays for banning fats from the kitchen. Nonetheless, fats are an indispensable part of the cook's repertoire. Let's consider the reasons why.

Deep frying, which involves temperatures of 200°C (392°F) or more, gives French fries and fritters their crispiness. Because water cannot withstand such temperatures without boiling, the surface of fried food is dried out without the water inside having time to diffuse outward. The crust that is formed in this way is what produces the sensation of crispiness. If deep-fried foods were cooked in a very hot oven, the results would be different. Chemists in Bristol and Nantes have demonstrated that fats are an essential element of Maillard reactions between sugars and amino acids. These reactions differ depending on the presence or absence of fatty matter; indeed, the good taste of the browned surface of fried foods is directly attributable to fats. It is for this reason that quail, for example, is wrapped in fat or bacon before roasting.

It is a mistake to baste meat with the juices that drip into the roasting pan, by the way; these juices are mostly water, which softens the crispy surface (itself the consequence of the water inside the skin evaporating) and so produces exactly the opposite result of what one hopes to achieve by roasting. Here again fat is the cook's friend. Ideally one would use a decanting drip pan to recover the melted fat while eliminating the water.

From Roasts to Emulsions

Whether one uses oil or butter, fats are inevitable in mayonnaise, béarnaise, hollandaise, beurre blanc, and other emulsified sauces formed from a base thickened with butter or cream. These emulsions consist almost exclusively of fatty matter. Oil droplets are packed together in water so closely that they no longer have room to move, with the result that the sauce has trouble flowing. In the case of some sauces the coagulation of the egg yolk adds solid particles to the oil droplets dispersed in the water, yielding suspension emulsions rather than emulsions proper.

Might it nonetheless be possible to increase the proportion of water and reduce the proportion of fat? One could use a hand mixer rather than a fork to divide the fatty matter up into smaller and therefore more numerous droplets, but little would be gained. Alternatively, one could use thickeners and gelatinizing agents, but it is difficult to reproduce the fluid behavior of an emulsified sauce in a suspension (of expanded starch granules, for example) or a concentrated solution (as when hydrocolloids—molecules surrounded by a lot of water molecules—are dissolved in water).

This observation may help us finally to transform a small evil into a great good. Butter, for example, acquires an unpleasant odor in the refrigerator because many aromatic molecules are soluble in fats. Makers of perfume exploit this solubility in order to extract fragrances from the most delicate flowers: They place freshly cut blossoms on a neutral fat for a few hours, then discard them and melt the fat in order to recover the aromatic molecules that have dissolved in it. This process is known as enfleurage. Butter serves as the fatty substrate for enfleurage in a refrigerator, as do chocolate (made of cocoa butter) and cream.

Separating Aromas

This property could be put to more systematic use in cooking. Why not wrap cheeses in aromatic plants, for example, so that the aromatic molecules slowly dissolve in the fatty matter of the cheese? We would also do well to recall the underlying principle of a famous recipe for sage butter, recommended in Italy as an accompaniment for pasta: When one cooks the leaves of this herb in butter, the heat causes their cells to burst and release aromatic molecules, which are then dissolved in the melted butter.

Not all aromatic molecules are fat soluble, however. One way to dissolve them is to use a separating funnel, long familiar to chemists as a useful device for separating mixtures. Put oil and water in the funnel, and then add chopped or ground pieces of an aromatic food such as cepe mushrooms. When the funnel is shaken, the hydrophobic aromatic molecules are dissolved in the oil while the hydrophilic aromatic molecules are dissolved in the water.

In this way two flavors are created out of one because the aromatic molecules are different in the two solvents. If you don't have a separating funnel, simply use a jar that can be hermetically sealed. Put oil and water in it, add an aromatic food, and when the aromatic molecules have been dissolved, slowly drain off the upper oily phase into another container, reserving the watery solution at the bottom.

How can these fragrant solutions be put to good use? If you were to prepare the cepe-scented water using egg whites instead of water, you could incorporate the cepe-scented oil by whisking it into the egg mixture. The egg proteins will then be tensioactive molecules, which will give you a cepe emulsion.

88

Mayonnaises

The art of mixing oil with water.

MAYONNAISE IS A REMARKABLE SAUCE: Through the miraculous intervention of the egg yolk, cooks manage to combine oil and water. Let's try our hand at making a few new varieties that preserve the spirit of traditional recipes while enriching our culinary arsenal.

What happens when one adds oil to a mixture of vinegar, mustard, egg yolk, and salt? To the naked eye the preparation is homogeneous, but a microscope shows that the ingredients are not thoroughly mixed together: Large oil droplets are dispersed in the small amount of water contributed by the vinegar, mustard (which is itself made with vinegar), and egg yolk. If the oil doesn't float on top but emulsifies instead, this is because the yolk's tensioactive molecules (one part of which is water soluble and the other fat soluble) coat the oil droplets, with the water-soluble part immersing itself in the water and the other part immersing itself in the oil. In egg yolks these molecules are primarily proteins and, in smaller proportion, lecithins.

Draped over the droplets of oil in this fashion, the tensioactive molecules prevent them from aggregating. These molecules cannot be seen under the microscope. Nor can the vinegar or the salt, which nonetheless play an important role, enabling the tensioactive molecules of the yolk to bring about an electrical repulsion between the oil droplets and thus helping to stabilize the sauce.

How much mayonnaise can be made from a single egg yolk? The amount of oil that can be added depends on two things: the quantity of water in which the oil droplets are distributed and the quantity of tensioactive molecules. A simple calculation shows that there are enough tensioactive molecules in a single egg yolk to make several liters of sauce if there is enough water because the oil content can be as high as 95%.

Why do mayonnaises go wrong when no more than a large bowlful has been made from one yolk? Because the oil droplets are so tightly packed together that they have trouble moving, and the sauce becomes firm. At this point, in order to prevent the sauce from breaking, it is necessary to add more water before adding more oil. In the case of homemade mayonnaise, by the way, use only a single drop of egg yolk. This is enough to coat all the droplets in your sauce.

A Yolkless Mayonnaise

Let's take matters a step further and do away with the egg yolk altogether. Once it is understood that a mayonnaise is an emulsion, which is to say a dispersion of oil droplets in water, one can experiment by modifying the ingredients. We may begin by replacing one set of tensioactive molecules with another. This type of molecule is found in many foods. The reason egg whites can be stiffened by whisking, for example, is that they are a solution of proteins that coat air bubbles.

Take an egg white and add to it a drop of vinegar, a little salt, and a little pepper. Now—slowly at first and then more rapidly—add oil while constantly whisking the mixture. A small amount of foam begins to form and then subsides as the oil is incorporated, exactly as in a mayonnaise. After having introduced a large quantity of oil, you have a yolkless mayonnaise.

An Eggless Mayonnaise

Can we push this idea still further? There is no particular reason to rely on the tensioactive molecules of an egg to give body to a mayonnaise. Why not use gelatin, for example? Its tensioactive properties are well known empirically to cooks, who use it to thicken a number of warm sauces. Let's melt some veal demi-glace (any highly gelatinous solution will do), add to it a bit of vinegar, a bit of salt if you like, and a bit of pepper, and whisk in the oil. Once again, the oil is very readily incorporated and forms what can be seen under the microscope to be an emulsion.

Is the result worthy of the name "mayonnaise"? To answer this question one has to decide exactly what a mayonnaise is. If it is a cold emulsion of oil in water, then these three preparations are all mayonnaises. If the taste of egg yolk is the thing that matters, then only the classic mayonnaise can legitimately be called by this name; indeed, there is a law in France that an emulsion can be called mayonnaise only if it contains more than 8% yolk. Recipes for mayonnaise vary widely, however. The great Antonin Carême, known in the nineteenth century as the Napoleon of the stoves (a compliment at the time), made his mayonnaise without mustard, using only olive oil from Aix, which he added with a wooden spoon. Other cooks swore by mustard. All tastes are in nature, after all.

The fact remains that an eggless mayonnaise, as we may provisionally call it, increases culinary freedom. Using mint jelly, for example, one may create a mint mayonnaise to accompany a traditional English leg of lamb. But one can also thicken a mayonnaise with various aromatic solutions—a rich lobster stock, for example, or an infusion of rosemary, thyme, and orange juice—to which gelatin has been added, either in powdered form or in leaves.

89

Aioli Generalized

Delicious emulsions that can be made from any vegetable, meat, or fish.

WE START WITH AIOLI—the real article made in Provence by adding olive oil to crushed garlic, without the benefit of egg yolk. It is an emulsion, which is to say a dispersion of oil droplets in water, supplied in this case by the garlic. Why should this sauce be stable, whereas normally a mixture of oil and water separates? Because garlic contains tensioactive molecules that coat the oil droplets and prevent them from fusing. Aioli is a relative of mayonnaise, in which the protein molecules and phospholipid lecithins of the egg yolk are tensioactive.

Let's try varying the traditional recipe a bit. Does the shallot, which belongs to the same vegetable family as garlic, also contain tensioactive molecules that permit us to make an "échalatoli"? Can an "oignoli" be made from onions? The experimental response is conclusive: Adding oil to crushed shallots or onions does in fact yield such emulsions. In the worst case it is necessary to add a little water (just as, in certain recipes for aioli, adding a piece of bread soaked in milk is recommended). It remains for cooks to dream up new dishes that these sauces could accompany.

But are there vegetables other than the ones of the lily family that could serve the same purpose? After all, cooks know that mustard can be used to make emulsioned vinaigrettes, for mustard also contains tensioactive molecules that help to stabilize the sauce.

The Virtues of Membranes

The answer is simple if we recall that all cells, whether plant or animal, contain compartments of water and proteins that are bounded by membranes. These membranes are composed of phospholipid molecules having a water-insoluble lipidic tail and a water-soluble head. In living cells these phospholipid molecules form double layers because the hydrophobic tails are grouped together inside the various cellular compartments, whereas their hydrophilic heads are in contact with the water outside the compartments. Phospholipid molecules are abundant and have strong tensioactive properties. Can they be used to thicken an emulsion?

Let's try crushing a zucchini and adding oil to it, drop by drop. The result is a thick sauce that, by analogy once again, we may call "courgettoli." Why not move on next from the plant to the animal kingdom, because animal cells also contain membranes? If we crush a cube of beef and add some oil, we find that we obtain "bœufoli." From any vegetable or animal, then, we discover that we can make an emulsified sauce as long as we release the phospholipid molecules, for example by crushing them and adding oil (again, a bit of water must be added at the outset if the chosen ingredients contain too little of it).

Phospholipids are not the only tensioactive molecules in plant and animal cells; many proteins have good emulsifying properties as well. Whisking oil into an egg white forms a remarkably stable emulsion but with very little flavor. More interesting sauces can be made from meat, fish, and vegetables, whose protein molecules likewise contribute to their stability.

New Mousses

From emulsions we turn now to an analogous physical system, foams, which consist of air bubbles dispersed in a liquid or solid. The fact that a chocolate mousse can be obtained by whisking a chocolate emulsion, as we shall see later in this section, suggests that it may be possible to generalize this procedure.

We might try using cheese, for example. Cheese contains a great amount of fat and tensioactive molecules, or caseins, which disperse the fatty droplets throughout the milk. To produce a cheese mousse, first make a cheese emulsion, heating a pan and then adding water and small rounds of goat cheese

(Chavignol, for example). Shake the pan gently. The result is a smooth, thick sauce that under the microscope can be seen to be composed of droplets dispersed in water (for a sharper taste one can substitute vinegar for the water and reduce it over high heat). This preparation we may call a "cheese béarnaise."

To obtain the desired mousse, use the same procedure that turns cream (an emulsion of milk fats in water) into whipped cream: cooling followed by whisking. In this case you put the pan in the freezer for about ten minutes and then whisk the chilled sauce vigorously to make what might be called a Chavignol Chantilly. The same method can successfully be used with Roquefort to yield a Roquefort Chantilly, and so on.

90

Orders of Magnitude

Dispersed systems make it possible to get a lot from a little.

IN MAKING FLANS ONE OFTEN SEEKS, for economic or dietary reasons, to limit the proportion of egg or increase the proportion of water. How far can one go in this direction? Much farther, certainly, than traditional cuisine has yet gone. Let's begin by analyzing two related cases: making mayonnaise from a single egg yolk and making meringue from a single egg white.

Cookbooks often say that one egg yolk is enough to yield a large bowl of mayonnaise. But in fact the physical chemistry of mayonnaise makes it possible, as we have already seen, to obtain liters of sauce from a single yolk: The quantity of tensioactive molecules in an egg yolk (proteins and phospholipids, equal to about 5 grams) is sufficient to cover a football field with a monomolecular layer; and when these molecules cover oil droplets whose radius in on the order of a micrometer (a millionth of a millimeter), as in the case of mayonnaise, they suffice to stabilize several liters of sauce.

We have also seen that classic mayonnaises break because they do not have enough water to sustain the dispersion of oil droplets. Verifying this claim experimentally is a simple matter, but it means wasting a lot of oil. Instead try making a large bowl of mayonnaise using only a single drop of egg yolk (the source of the tensioactive material) dissolved in a teaspoon of water (or vinegar if you want to taste the result).

A Cubic Meter of Foam

A stiffly whipped egg white is essentially the same thing, only the oil in mayonnaise (which is insoluble in water) has been replaced by air (also largely water insoluble), and the tensioactive molecules that isolate the air are now proteins, which make up 10% of the egg white. The result is no longer an emulsion but a foam. And so our question remains essentially the same: How much meringue can be obtained from a single egg white?

Here again, calculating orders of magnitude is simple. One begins by estimating the number of protein molecules in the egg white. Then one determines the total surface that can be occupied by these proteins and the number of air bubbles that can be covered by these proteins. Multiplying this number by the volume of a bubble, one obtains the maximum volume of meringue that can be made from a single white: several liters, indeed several cubic meters, depending on what assumptions you make about the size of the bubbles and the type of protein coating.

Why, then, does one generally wind up with only a small cubic decimeter of foam? There's no shortage of air, so the problem must once again be a lack of water. If you add some water to an egg white and whisk you will find that the volume of the foam increases, and if you keep on beating the mixture vigorously you will eventually end up with several liters of meringue. This version is less stable than the classic preparation, however. The stability of a foam depends on the viscosity of the liquid (which is reduced by the addition of water) and the size of the bubbles (which determines the action of capillary forces between the bubbles).

Extreme Flan

The fact that emulsions and foams are both dispersed systems suggests that other such systems found in cooking may display the same behavior. Consider what happens in the case of a quiche. First one lines a baking pan with dough and fills it with cubes of bacon and a mixture of eggs and milk. Yet the cook who skimps on eggs (perhaps because they cost more than the milk) sometimes ends up adding so much milk that the quiche remains liquid after cooking. Because milk is mostly water, the question now becomes: How much water can we add to an egg and still obtain the equivalent of a flan?

Let's suppose that the proteins responsible for coagulation are spherical. In the course of coagulating they can be assembled either in a compact fashion, so that no water is trapped between them; or next to one another along the edges of a very large cube that encloses a maximum amount of water; or along the edges of a network that, for the sake of simplicity, may be assumed to be cubic.

In the first case, the volume of water enclosed is zero. In the second case, a simple calculation shows that the volume would be impossibly large, on the order of several cubic meters. Let us therefore consider an intermediate case, based on observation of a fried egg using an atomic resolution microscope. This time, an order-of-magnitude calculation leads to a volume of flan—what physical chemists call a gel—equal to a liter.

To check this calculation experimentally, take an egg and add to it an equal volume of water, and then heat the mixture. Having successfully obtained a flan by this method, repeat the experiment while progressively increasing the quantity of water: twice the original volume, three times, and so on. You will discover that almost three-quarters of a liter of water can be made to "gelatinize" with a single egg—the same order of magnitude predicted by the calculation.

The number of different types of dispersed systems is huge. So far we have explored a liquid dispersed in a liquid, a gas dispersed in a liquid, and a solid dispersed in a liquid. But other combinations remain to be investigated.

91

Hundred-Year-Old Eggs

Experiments with acids and bases.

EGGS USED TO BE PUT IN SAND, SAWDUST, OR WAX in order to be preserved. Asian peoples devised recipes that took advantage of aging, instead of compensating for its effects, in order to create what were variously known as hundred-year-old eggs, centenary eggs, and even thousand-year-old eggs—names that symbolized links with the past as well as longevity. How credible are these recipes from the chemical point of view? A few experiments reveal the unexpected behavior of eggs in acid and basic environments.

The origins of the Chinese art of preserving eggs are lost in the mists of time. Initially, eggs seem to have been immersed in juices extracted from a local tree. Then it was discovered that by putting them in a mixture of ashes and earth and keeping them in a dark, cool place one could obtain the same culinary result but in only ten to twelve weeks. What inspired these practices? Can others be imagined?

Recipes for hundred-year-old eggs vary from region to region in China. Some call instead for placing duck eggs in a plaster that contains various ingredients: lime, saltpeter, bicarbonate of soda, mud, fragrant herbs, tea, rice straw, and so on. The eggs are left to rest for at least three months, and their flavor is said progressively to improve. It is surprising to note that some of these ingredients are also used in parts of France; even more recent civilizations such as our own make use of lime and ashes, which contain potash (potassium hydroxide). Thus recipes for preserving eggs fall into two classes: ones that contain only eggs and others that put them in contact with an alkaline compound. What is the effect of these bases? The effect of acids? After all, at least one modern French recipe also advises placing quail eggs in vinegar.

An Egg in Vinegar

Let's experiment by placing a whole egg, in its shell, in a large transparent container. When it is covered with white vinegar, bubbles soon escape from the shell. Why? Because the acetic acid of the vinegar is attacking the calcium carbonate? A lighted candle, placed in the container, eventually goes out, a sign that the acid gives off carbon dioxide, which, being denser than air, accumulates in the container, driving out the air (the same thing can be demonstrated more technically by collecting the gas in limewater, which becomes cloudy). Then, after half a day or so, a thin, red surface layer detaches itself from the shell. This is why the eggs have a pink shell: the white of the carbonate and the red of this layer combine to produce the final color (at least in the case of French eggs; I have heard that eggs in England remain white).

Let's continue observing. After one or two days of slow gaseous emission, the egg seems to have gotten bigger. Is this merely an illusion? The sequence of events shows that the enlargement is real: The final volume can be more than twice the initial volume. The shell has been completely dissolved, but the contents of the egg have not spread into the vinegar, for the acidity causes the white to coagulate. Experiments with several eggs allowed to sit for different periods of time show that this coagulation, limited at first to a thin outer layer, extends to an increasing proportion of the albumen, reaching even as far as the yolk. On reflection, this effect is not entirely surprising, for one finds the same thing when one pours vinegar on an egg white in a bowl: The superficial layer of the white coagulates, among other reasons because the H^+ ions contributed by the acid prevent the acid groups of the proteins from being ionized while triggering the ionization of the base groups, which thus become positively charged (bases have the opposite effect). Electrical repulsions between the charged groups of proteins thus unfold the proteins, which are then bound by forces called disulfide bridges that link sulfur atoms. Coagulation occurs because water is trapped by the resulting network of proteins.

Osmotic Expansion

If the dissolution of the shell and the coagulation of the albumen are simply explained, it may seem less clear why dilation occurs. Could it be that osmosis is responsible for the increase in size? Water molecules tend to go from areas where they are most concentrated to areas of least concentration. Whereas the water concentration is about 95% in the vinegar, it reaches only 90% in the egg white. Moreover, whereas the acetic acid migrates toward the interior of the white (this can be verified by measuring the acidity of an egg white that has sat for several weeks in vinegar), the protein molecules dissolved in the water of the whites are too large to pass through the coagulated membrane. In other words, the water of the vinegar enters into the albumen, increasing the water concentration inside the egg.

To show that this is what happens, one has only to leave the eggs in an acetic acid solution whose acid concentration is greater than 10%. Once again the shell is dissolved, but this time the egg ends up being smaller because the osmosis is reduced. How would an egg placed in a concentrated solution of chlorhydric acid turn out? One would obtain the same smaller egg produced by the acetic acid solution, but the coagulation would be more rapid and clearer.

The Floating Yolk

I invite you to conduct your own experiments; many other surprises await those who are patient enough to observe carefully. For example, when the shell is dissolved and the egg white is still translucent, you can actually see the yolk floating in the white.

As for bases, adding caustic soda (sodium hydroxide) to an egg white causes it initially to coagulate. A chemical reaction produces a nauseating sulfur gas, and the egg then turns clear again. Obviously the soda dissociates the proteins after having first precipitated them. If we put eggs in ashes or in lime, which have lower pH levels, we can wait—for a hundred years.

Smoking Salmon

Sugar and an electrical field can be used to accelerate smoking.

SMOKED SALMON IS AN EXPENSIVE DELICACY that France has long specialized in producing. Manufacturers buy imported salmon and resell their smoked filets the world over. A team of researchers from the Institut Français de Recherche pour l'Exploration de la Mer and the Centre de Coopération Internationale en Recherche Agronomique pour le Développement perfected the process currently used to accelerate processing without sacrificing flavor. The two principal ingredients of the new method are osmosis and electrostatic smoking.

Smoking, like salting and drying, was originally used to preserve foods. In all three cases the idea was to eliminate water from foods in order to kill microorganisms that were already present and to prevent the development of new pathogenic microorganisms. Yet the old methods often gave an excessively salty or smoky flavor.

With the appearance of modern refrigeration systems, the technique of smoking was retained, in a modified form, because it gave filets of fish (and other foods) the delicious taste we know and love. Present-day products are less salty and less smoky, but they must be kept at low temperatures, between 0°C (32°F) and 2°C (36°F).

Fish filets prepared by traditional methods are either immersed in a brine, sprinkled with salt, or injected with brine. The first treatment, which takes about four hours, eliminates only about 2% of the water; health regulations make it a complicated affair requiring costly treatment facilities. The elimination of water is ensured only by drying at a temperature of about 22°C (72°F), with a humidity of 65%, for three to four hours, before the actual smoking begins. Here again the procedure must be carefully monitored, for the processing temperatures favor the development of microorganisms.

In the procedure patented by Antoine Collignan, Camille Knockaert, Anne-Lucie Wack, and Jean-Luc Vallet, the salting and drying are done simultaneously at a temperature of 2°C (36°F). The filets are immersed in a concentrated salt and sugar solution (10–30% sugar and 70–90% salt) so that some of the salt penetrates the flesh by osmosis and dries it out. In fact, the various molecules are distributed in such a way that their concentration is everywhere the same: When the filets are immersed in a salt-saturated solution, the salt migrates toward the flesh and the water comes out. At the same time the sugar—composed of large molecules that cannot enter the cells of the fish—promotes the outflow of water as well. All told, filets immersed in a solution containing some 350 grams of salt per liter and about 1900 grams of sugar per liter lose roughly 10% of their water. Moreover, the sugar reacts with the amino acids of the fish and produces agreeable flavors through a series of Maillard reactions. In certain classic recipes for smoked salmon the fish is rubbed with sugar until it has a tanned appearance. With the new procedure, the result is comparable but easier to achieve because it takes place in solution.

Smoke Without Fire

After rapid rinsing and draining, the smoking takes place in a chamber traversed by a metal conveyor belt that passes under a grate. Smoke is produced by subjecting sawdust to pyrolysis (dry heat). The smoke is injected into the chamber after having been cooled to a temperature of 40°C (104°F), condensing the aromatic polycyclic hydrocarbon molecules, which are carcinogenic, so that they do not contaminate the filets. After being electrically charged as they pass under the grate, the smoke particles are impressed on the filets by a difference in electrical potential (amounting to several tens of thousands of volts) between the conveyor belt and the grate. In this method the smoking lasts fifteen minutes (rather than three and a half hours, as in the usual procedure) so no additional drying out occurs.

The electrical system could easily be adapted to the scale of traditional smoking, where up to 400 kilograms (880 pounds) of salmon filets can be treated at a time, and the new method of salt drying and cold smoking could be applied to various other meats and fish. Naturally, one wants to know whether products prepared in this way are as good as the old ones. At first their coloring was a bit more pronounced than that of smoked filets today, but by prefiltering the smoke it became possible to give them a lighter tint. As for the flavor, it is not noticeably different from that of salmon smoked in the traditional manner.

93

Methods and Principles

On the invention of new recipes.

WE COOK TODAY THE WAY PEOPLE COOKED in the Middle Ages, content to mechanically execute fixed recipes—this at a time when space probes are being sent to Mars. We need to ask ourselves how reflection and rationality can be combined to renew creativity in cooking.

Certain types of cooking are pure in the sense that they involve only a single physical phenomenon. The oldest ones are those in which heat is transmitted through conduction. Since ancient times only the materials that transport the heat have changed.

The first type of pure cooking consists of putting food in direct contact with a hot solid: on stones heated by embers, for example, or on a cast iron plate heated by fire, inside a layer of salt that is heated, or in a mold that is heated by fire or placed in an oven (as in the cooking of custards, flans, and so on).

Heat can be transmitted by a hot liquid as well. If this liquid is boiling water, one can use it to boil meats; if it is simmering water (or sauce), one obtains various poached dishes (fricassees, blanquettes, matelotes, fillings for vol-au-vents, and so on).

Cooking can also be done by means of hot air. In the case of roasting, the air is dry and heated to a temperature greater than 100°C (212°F). In the case of drying or smoking, the air is dry and the temperature less than 100°C (212°F). In the case of braising, steaming, and cooking *en papillote* and *en croûte,* the air is moist.

TABLE 1. DOUBLE-COOKING METHODS

First this cooking method →

Then this one ↓

	Contact with Solid	Simmering Water	Boiling Water	Warm, Dry Air	Hot, Dry Air	Humid Air	Oil	Infrared	Microwave	Acidification
Contact with Solid	1	2	3	4	5	6	7	8	9	10
Simmering Water	11	12	13	14	15	16	17	18	19	20
Boiling Water	21	22	23	24	25	26	27	28	29	30
Warm, Dry Air	31	32	33	34	35	36	37	38	39	40
Hot, Dry Air	41	42	43	44	45	46	47	48	49	50
Humid Air	51	52	53	54	55	56	57	58	59	60
Oil	61	62	63	64	65	66	67	68	69	70
Infrared	71	72	73	74	75	76	77	78	79	80
Microwave	81	82	83	84	85	86	87	88	89	90
Acidification	91	92	93	94	95	96	97	98	99	100

Double-cooking methods are numerous; each cell in this table represents one possibility. For example, cell 25 represents cooking first in hot, dry air, then cooking in boiling water.

Classic French cuisine also involves the indirect transmission of heat by means of infrared rays, most notably in the case of *rôtissage à l'ancienne* (because in true roasting the meat must be placed in front of the heat source, not above it). A laser or, more generally, visible high-energy waves can be used for the same purpose, with similar results. A well-known innovation that allows heat to be transferred directly to foods is microwave cooking.

Finally, acid is used by some peoples as a medium for cooking fish, as in Tahiti and Central and South America (seviche).

Double Cooking Methods

Classical cuisine sometimes superimposes or otherwise combines these pure types. For example, grilled meats sometimes are the result of heating first by radiation and then by means of a dry fluid (typically air, but in principle any other gas). Braising in the traditional style is done first by browning the outside of meats in a hot oven and then simmering them in a liquid.

Are other combinations possible? It may be useful to follow the example of Dimitri Mendeleev, who sought to make sense of the apparent disorder of chemical elements by constructing a table in which elements are presented in the order of their atomic mass in rows and grouped in columns according to the similarity of their chemical properties. In the same spirit one can construct a table whose entries along both rows and columns are pure types of cooking. The numbered cell at the intersection of each row and column designates the type of dish that results from first cooking a food in the manner indicated at the head of the column and then cooking it in the manner associated with the row.

A number of cells in this table identify familiar culinary techniques. For example, cell 20 (acidification followed by simmering in liquid) corresponds to the classic recipe for cooking wild boar. Cell 63 (boiling in water followed by frying in oil) corresponds to certain recipes for French fries. Other squares refer to unknown procedures, however. Additional lines and rows could be incorporated to take account of novel techniques such as cooking under very high pressures.

Expected Inventions

Surely new procedures can be invented. In certain cases the procedure suggested by mechanically applying the methods indicated seems unpromising. For example, why should cooking be done twice in contact with a solid (cell 1)? Nonetheless, the cells lying along the principal diagonal that runs from the upper left corner to the bottom right are not uninteresting; for instance, the two fryings associated with cell 67 correspond to the method of deep-frying potatoes by two immersions in hot oil.

Elsewhere novel procedures must be devised. For example, cell 27 involves frying first and then cooking in boiling water. The frying would produce a dry and hardened surface layer in addition to specific flavors. To be sure, the subsequent cooking in boiling water would destroy the crispiness achieved by this method, but it would also redistribute the aromatic molecules created during the first phase. The overall flavor therefore would be quite different from the one obtained by braising.

This is only one example. Other squares suggest new possibilities and invite inspired chefs to travel the paths they open up for exploration. Once these combinations have been charted, the next task will be to construct tables in more than two dimensions.

94
Pure Beef

A textural additive for restructuring meats.

WHY DO THE FRENCH EAT LESS BEEF than they did only ten years ago? In part it is because the quality of the meat does not justify its high price. Tender cuts that can be cooked rapidly (various types of steak, for example) are expensive because they constitute only 20–30% of bovine muscle tissue. What can be done with the other cuts that used to be tenderized by long, slow cooking?

The meat processing industry has proposed a solution in the form of restructured products such as ground beef and meats sliced into thin sheets. Both techniques—cutting fresh meats into very small pieces and cold-hardened meats into very thin slices—destroy the collagen networks that form the majority of the muscles in the anterior half of the animal. Yet once it has been destructured, meat must be put back together: The butcher in your supermarket reshapes ground beef with a press, for example, but there isn't always much left of the reformed product after cooking.

Cooks know that cooking an egg white produces a gel that can trap dispersed particles; this is the principle of fish loaves, clafloutis (a custard made with cherries or plums), and quiches. Drawing on this principle in order to restore cohesiveness to destructured meats, the processing industry has introduced various textural additives with increased binding power, such as sodium alginate, an extract of algae. However, the use of dairy and vegetable compounds makes it impossible to advertise the product as pure beef.

Many researchers have looked for ways to extract such additives from meat itself, particularly from the scraps that remain on the bones after butchering. Teams led by Joseph Culioli and Ahmed Ouali at the Institut National de la Recherche Agronomique (INRA) stations in the Clermont-Ferrand area have shown that, at least in the laboratory, myosin can be used as an effective binding agent.

Myosin is an abundant protein, accounting for 20% of the dry matter in the striated skeletal muscles that can be processed by mechanical means. Proteins in the muscle mass of the animal fall into three categories: myofibrillary proteins, sarcoplasmic proteins, and connective tissue proteins. Myosin is the principal protein of the myofibrillary system, where it is found in the form of thick fibers. In the presence of calcium ions and adenosine triphosphate (ATP), the fuel of living cells, these fibers combine with more delicate fibers composed mainly of actin, giving rise to muscle contraction when the two sorts of fiber slide past one another.

Myosin Gels

Because the properties of proteins depend on their amino acid sequences, they do not all have the same gelatinizing effect. The thermal gelatinizing properties of myosin, which are also involved in the preparation of cooked hams, pâtés, sausages, and so on, are greater than those of actin. To determine which combination of factors yields the firmest gel, the INRA biologists developed new extraction methods to compare the effect of myosin proteins from two different types of muscle (fast white muscles, which are responsible for brief spurts of intense effort, and slow red muscles, which function in the presence of oxygen) removed from animals at different times after death. Initially myosin was extracted from muscles in rabbits because they are unmixed, being either red or white.

Protein samples were extracted by grinding up the muscles and then placing them in solutions with different concentrations of salt. The thermal gelatinization of the protein suspensions in these solutions was studied with the aid of a rheometer (an instrument for measuring the flow of fluids), which revealed both the viscous and elastic characteristics of the gels.

A Hot Gel

These measurements showed that thermal gelatinization occurs even at very weak myosin concentrations (0.1–0.5%) and that the firmness of the resulting gels strongly increases with the myosin concentration.

The gelatinization of pure myosin begins at 40°C (104°F), with the rigidity of the gels increasing up to 80°C (176°F). The purer the myosin solution, the greater the rigidity. As with other gels, the rigidity of myosin gels depends on the salt concentration and the acidity of its environment. The myosins that form the firmest gels are found in the fast white muscles, with a pH of about 5.8, and in the presence of salt, which favors the dissociation of macromolecular chains.

Subsequent analysis confirmed the suitability of bovine myosin for the restructuring of meat. The gels are firmer when the myosin is extracted just after slaughter, before the onset of rigor mortis (a consequence of irreversible bonds being established between the actin and myosin). Sodium pyrophosphate, a molecule belonging to the same family as ATP, dissociates the actin–myosin complexes and so makes it possible to obtain firmer gels. The incorporation of these myosin extracts in meats that are then sliced into thin sheets has made it possible to increase their cohesion while limiting the loss of juices, which are trapped in the gel.

95
Fortified Cheeses

The right bacteria can strengthen the flavor of cheeses.

A GOUDA OR A CHEDDAR TAKES ON its full gustatory quality only after several months, and many cheeses, even ones that have been aged for a long time, do not have the powerful flavor that one might want. The aging of cheese has been a lively topic of debate in the gastronomic world for centuries. The milk that is curdled and seeded with lactic bacteria acidifies in the course of maturing: The transformation of the milk sugar lactose into lactic acid prevents contamination by pathogenic microorganisms, and the lactic bacteria release aromatic compounds that contribute to the taste of the cheese. Mireille Yvon and her colleagues at the Institut National de la Recherche Agronomique (INRA) station in Jouy-en-Josas have studied which strains of lactic bacteria produce the greatest quantity of these aromatic compounds.

The INRA biochemists knew that the flavor patiently acquired by cheeses results from the microorganisms responsible for maturing. Fats and sugars are progressively transformed, and proteins are dissociated into their constituent elements, amino acids, which are then transformed into aromatic molecules. For example, the amino acids leucine and valine produce compounds having a cheese note, whereas phenylalanine, tyrosine, and tryptophan are the precursors of floral and phenolic notes (as they are described by trained tasters).

Does the slow dissociation of proteins into amino acids limit the formation of aromatic compounds? No, for the direct addition of free amino acids does not improve the taste of cheeses any more than does the seeding of milk by lactic bacteria (whose capacity for dissociating proteins has been increased by genetic engineering). In the latter case, the amino acids are released in greater quantities, but the taste is not changed. In the early 1990s, Yvon and her team concluded instead that what limits the development of flavor is the transformation of amino acids into aromatic compounds.

Stimulating Additions

Two biochemists in the Netherlands, W. Engels and S. Visser at Wageningen University, noticed that the flavors typical of Gouda were obtained when methionine was added to lactic bacteria (without milk) and went on to identify two enzymes that seemed to be responsible for the phenomenon. Yvon and her colleagues, for their part, had observed in vitro that lactic bacteria degrade certain amino acids to form aromatic compounds such as aldehydes and carboxylic acids. The first step in this transformation, known as transamination, involves a reaction between an amino acid and a molecule named ketoglutarate, which produces a keto acid in addition to glutamate (a molecule that, as we have seen, is used as a flavor enhancer in Asian cuisine and in many commercial products because it communicates the taste that we know today as umami). During transamination, an amine group ($-NH_2$) is converted from an amino acid to a keto acid, which is then chemically modified and transformed into aromatic compounds.

In 1997, the biochemists at Jouy succeeded in purifying and characterizing aminotransferase, the enzyme in the lactic bacterium *Lactococcus lactis,* which is responsible for the transamination of leucine and methionine. Nonetheless, under actual aging conditions, the presence of this enzyme does not significantly affect the aromatic quality of cheeses. Why? Was the diffusion of reactive agents in the lactic bacteria too slow? Did the lactic bacteria lack the molecules necessary for them to receive the amine groups?

The INRA researchers tested the second hypothesis by seeding warm (pasteurized) milk with lactic bacteria and then adding rennet (which curdles the milk and transforms it into cheese). In this way they obtained a curd that they then molded and pressed and finally immersed in a brine enriched with ketoglutarate. They followed the transformation of the amino acids during the aging process, and a panel of tasters analyzed the development of the cheese's odor.

In the control cheeses, which had not been enriched by ketoglutarate, very few amino acids were dissociated, and the odor was weak. By contrast, the addition of ketoglutarate augmented the transformation of several amino acids. The transformation of ketoglutarate led to the formation of powerfully aromatic compounds, such as isovalerate in the case of leucine and benzaldehyde in the case of phenylalanine, demonstrating that the odor of the cheeses was increased by the addition of ketoglutarate. Similar results were obtained for Cheddars, tested for comparative purposes.

In Search of Efficient Microorganisms

While they were studying the effects of adding ketoglutarate, the INRA biochemists observed that the glutamate produced during transamination is transformed by an enzyme produced by other bacteria in ketoglutarate. Because glutamate is abundant in milk, even before aging, they had the idea of introducing the gene for dehydrogenase glutamate, the enzyme they had discovered in a lactic bacterium, which they suspected would produce ketoglutarate from glutamate.

The effects of introducing this gene were followed in vitro and in a control cheese. The modified lactic bacteria were found to trigger the transamination of the amino acids no less completely than lactic bacteria to which ketoglutarate had been added. What is more, lactic bacteria containing the dehydrogenase glutamate gene produced more highly aromatic carboxylic acids.

It is clear, then, that fortified bacteria can be used to make better cheeses. The problem is that genetically modified organisms are not universally accepted by consumers. Therefore biochemists are searching for strains of lactic bacteria that naturally produce dehydrogenase glutamate. In this case, at least, genetically modified organisms will have served as a research tool. Is this not one of their chief advantages?

96

Chantilly Chocolate

How to make a chocolate mousse without eggs.

THE WORDS *CHANTILLY CREAM* conjure up images of fresh strawberries, ice cream, and airy desserts. Chantilly is a kind of foam, or mousse, made by whipping cream in a chilled bowl. When the whisk is guided in a circular motion, through a vertical plane, its wire loops steadily introduce air bubbles in the cream that are stabilized by the molecules of the casein (a protein) and by the crystallization of the fatty droplets. This crystallization takes place at a low temperature, which is why the cream and the bowl must be chilled beforehand. This cooling process also prevents the cream from turning into butter. To obtain the best results, stop whipping the cream once strands begin to form inside the loops of the whisk.

Can the fundamental principle of Chantilly cream be applied to fatty matter other than milk? Because chocolate contains cocoa butter, for example, it ought to be possible to make Chantilly chocolate.

A Chocolate Emulsion

Our chances of obtaining such a foam will increase if we begin by creating a physicochemical system similar to cream but with a chocolate base. Physical chemists know that cream is an emulsion, a dispersion of fatty droplets in water (in this case the water contained in milk, which also dissolves sugars, such as lactose, and mineral salts, but these ingredients, although they contribute to the taste of Chantilly cream, are unimportant for our purposes here).

The fatty droplets in an oil-in-water emulsion such as cream do not combine with one another, for they are stabilized by casein micelles and calcium phosphate. The casein molecules are bound together by the calcium phosphate into tensioactive structures, or structures with a hydrophobic tail immersed in the fatty droplet phase and a hydrophilic head immersed in the water phase.

The emulsion we need to make Chantilly chocolate can be formed in an analogous manner by mixing together water, tensioactive molecules, and cocoa butter. One simply pours a little water into a pan (which will be improved from the gastronomic point of view if it is flavored with orange juice, for example, or cassis purée) and adds some tensioactive molecules, either proteins from the yolk or white of an egg or gelatin (often used to thicken butter and cream sauces, which are also emulsions). One could rely simply on the lecithin already present in chocolate, but let's use gelatin instead and dissolve it in the water by heating. Then whisk in the chocolate. The result is a homogeneous sauce—precisely the chocolate emulsion we were looking to create.

From Emulsion to Foam

With this emulsion we can make a foam. Put the pan in a bowl partly filled with ice cubes to crystallize the chocolate around the air bubbles that we will next introduce by whisking the chilled sauce, either manually or with an electric mixer. The procedure is then exactly the same one followed in the case of Chantilly cream. Whisking creates large air bubbles in the sauce, which steadily thickens. Once the crystallization temperature is reached, the volume of the sauce suddenly expands, and its color changes from dark to blond chestnut.

This lighter color results from the air bubbles, which can be seen under a microscope. They also gradually change the texture of the sauce: After a while strands of chocolate form inside the loops of the whisk, just as in the case of Chantilly cream. In this way one obtains a foam that, unlike classic versions, is unadulterated by crème fraîche or stiffly whipped egg white. It is a purely chocolate mousse!

Want to try? Melt a half pound of chocolate with about 6 ounces of water. Three things can go wrong. If your chocolate doesn't contain enough fat, melt the mixture again, add some more chocolate, and then whisk it again. If the mousse is not light enough, melt the mixture again, add some water, and whisk it once more. If you whisk it too much, so that it becomes grainy, this means that the foam has turned into an emulsion. In that case simply melt the mixture and whisk it again, adding nothing.

97
Everything Chocolate

How to introduce chocolate into all kinds of pastry.

AT CHRISTMAS AND ON NEW YEAR'S EVE, chocolate is obligatory. But in what form? Chocolate puff pastry, perhaps? Chocoholics know that chocolate contains cocoa butter, and they would like nothing better than to be able to substitute it for ordinary butter in puff pastry. But they also know that the hardness of chocolate stands in their way. A few simple observations about state transitions will make it possible to solve this problem and to adapt the majority of classic recipes for pastry to new uses.

To make puff pastry one first makes a paste by kneading flour with a little water, sometimes butter. Next one rolls out the dough and places a layer of softened butter over it. The edges of the dough are then folded back over the butter so that it is completely covered. This envelope is then folded and rolled out six times, with the result that the dough that finally goes in the oven is composed of hundreds of alternating layers of dough and butter.

How can we incorporate chocolate in the dough? Dark chocolate cannot be used in place of butter, despite its cocoa butter content, because it is too hard. Eighty percent of cocoa butter—the only fatty matter permitted by law in France to be used in making chocolate, although other kinds have been proposed—is composed of three triglycerides, or molecules made up of glycerol (commonly called glycerine). These molecules are associated with three fatty acids: palmitic acid, stearic acid, and oleic acid.

Controlling Fusion

This composition explains the remarkable physical properties of cocoa butter. If cocoa butter were composed of only one sort of molecule it would melt at a fixed temperature, just as frozen water melts at 0°C (32°F) under normal pressure. But because it is formed of several types of molecule, its fusion point extends over a range of temperatures from −7°C (19°F) for some of the triglycerides to 34°C (93°F). Even so, 75% of the constituents of cocoa butter melt between 20°C and 34°C (68°F and 93°F) and almost 50% between 30°C and 34°C (86°F and 93°F). In other words, although chocolate is not a pure body, it is not very far from being one.

This characteristic, which is an advantage in eating chocolate bars, is troublesome for the cook, who generally uses it at temperatures in the neighborhood of only 20°C (68°F), hence the difficulties involved in using it in puff pastry, in particular. Comparing cocoa butter with ordinary butter, a more heterogeneous substance, suggests a way to overcome them: The less pure the chocolate, the more malleable it will be. This isn't a new idea; cooks have long been accustomed to melting chocolate with butter in order to obtain softer chocolate preparations. Nonetheless, one can take the idea farther by heating a neutral oil and adding chocolate to it. The chocolate mixes perfectly with the oil, which modifies the fusion properties of the chocolate in the desired manner.

Not only does this method enable us to make chocolate puff pastry (another solution, by the way, would have been to add cocoa powder to flour or to ordinary butter), the underlying principle makes it possible to reconfigure recipes for all kinds of desserts.

Make All Desserts with Chocolate

From puff pastry it is but a short step to make both shortcrust pastry (including the basic pie dough used to line meat or fish pies) and sugar crust with chocolate; one has only to replace butter with chocolate whose lipid composition has been changed. Similarly, one could make chocolate savarin dough, chocolate brioche dough, chocolate cream puff dough, almond or chocolate cookie dough, and so on. One might even be tempted to use chocolate in an almond custard: Add almond powder and an alcohol to melted chocolate, and then fold in the crème pâtissière (made by cooking egg yolks, sugar, milk, and flour).

Where the fatty matter used is cream rather than butter, it is the cream that must be replaced by a chocolate preparation. In this case it is not enough to manipulate the melting point of the chocolate, for cream is above all an emulsion, a dispersion of fatty droplets in water (from the milk). The droplets remain separated from one another for a long time, though not indefinitely, because they are surrounded by tensioactive molecules, one part of which is hydrophilic (immersed in the water) and another part hydrophobic (immersed in the oil). To replace the cream you will have to make a chocolate emulsion, which is not difficult. In a pan, heat water or a watery solution (coffee, tea, Cognac) together with squares of chocolate. The resulting "chocolate béarnaise" is chemically the equivalent of cream.

Finally, to make a chocolate bavarois (Bavarian cream), a new procedure is necessary. We begin by following the classic recipe to make a crème anglaise: Whisk powdered sugar and egg yolks together, add milk to the mixture, and cook it, then dissolve gelatin in the custard and fold in whipped cream. The problem is how to incorporate the chocolate. It cannot be substituted for the egg yolks, for it is their coagulation that gives the crème anglaise its texture, as particles of the cooked egg are suspended in the water of the milk. Nor can it be substituted for the gelatin.

How about replacing the cream with chocolate? No; the recipe calls for whipped cream. Forget the recipe! Whip up an emulsion of chocolate instead, just as we did in the last chapter to make our Chantilly chocolate, and add this mousse to the final preparation.

98

Playing with Texture

Gelatinizing emulsions produces a new kind of chocolate cake.

EMULSIONS ARE AN INEXHAUSTIBLE SOURCE of culinary discoveries. Here we will use them only as a point of departure for investigating more complex physicochemical systems that anyone can use in cooking.

In the preceding chapters we have considered several types of emulsion, foremost among them mayonnaise, the prototype described by all textbooks dealing with the physics of soft matter. Mayonnaise is a dispersion in water of oil droplets stabilized by the proteins of the egg yolk. A great many variations on this theme are possible. We have already looked at two of them: one made without a yolk, the other without any egg at all.

A yolkless mayonnaise is made in the same way as a classic mayonnaise, but in place of the yolk one uses the white of the egg. Albumen is made up of 90% water and 10% proteins, which have the same type of tensioactive properties as the proteins of the yolk. In a bowl one adds oil to an egg white, drop by drop, while whisking. At first the albumen foams, but gradually the air in the bubbles is replaced by the oil, which comes to be divided into smaller and smaller droplets by the shearing action of the whisk. The oil droplets, like the bubbles, are stabilized by the proteins of the egg white, for the proteins are unfolded in the course of whisking, so that their hydrophobic parts come into contact with the oil while the hydrophilic parts remain immersed in the water.

An eggless mayonnaise may be obtained by dissolving a half-sheet of gelatin in a small amount of heated water (or, for example, stock made from shellfish) and then whisking oil into the liquid just as one does in the case of a mayonnaise. At first a white emulsion appears as the proteins contributed by the gelatin attach themselves to the oil droplets at the interface of the water and oil. Once this emulsion has settled and cooled, it is transformed into a gel as the gelatin molecules become linked together at their extremities, forming a network within which the emulsified liquid is trapped.

Gelatinized Emulsion

The process of gelatinization can be viewed indirectly under the microscope. The oil droplets partially coalesce because the gelatin molecules coating them migrate away from the water–oil interface, forming a network (or gel) that traps the enlarged droplets. Once the gel has formed, by the way, whisking it again causes it to break down (in a classic mayonnaise, by contrast, the gel is stabilized further by whisking). Beating the gel dissociates the bonds of the network, with the result that the oil droplets are released.

The principle of a gelatin-based mayonnaise, then, is that an emulsion is created and then trapped in a physical gel, which is to say a gel that breaks up on being heated and reforms on being cooled. Can we devise additional variations on this theme by changing certain ingredients? Let's go back to the mayonnaise made with egg whites and look for a way to chemically gelatinize it—in other words, to cook it. What we want to end up with is a chemical gel that, as in the case of gelatin itself, is permanent rather than physical.

Cooking Mayonnaise

Let's begin by cooking the egg white mayonnaise in a microwave oven for about a minute so that the proteins covering the oil droplets are fused. What we get is a coagulated mass in which the oil is trapped, which is to say a physical system similar to a gelatinized mayonnaise. But is the oil securely retained by this body? If you squeeze the gel you will see that the oil comes out.

By manipulating the texture of the mayonnaise we seem only to have created a sponge for soaking up oil. Let's try replacing the oil with chocolate (which is composed mostly of cocoa butter) and melting it in a pan with a bit of liquid that contains water (whether from rum, coffee, orange juice, or something else). Then, when the temperature of the resulting chocolate emulsion is still lower than the temperature at which albumen coagulates (62°C [144°F]), whisk the chocolate emulsion into an egg white. Finally, put this mixture in the microwave. What will happen? The proteins of the egg white will gelatinize and imprison the chocolate emulsion.

If you try this experiment yourself you will end up with a delicious chocolate cake whose flavor is much more powerful than that of an ordinary chocolate cake, probably because the chocolate is in a dispersed state, producing a novel texture (which you can vary as you like by modifying the proportions of water and chocolate). I suggest calling this dessert a chocolate dispersion. It remains to study how gelatinization disturbs the emulsion in the three systems we have considered.

99 Christmas Recipes

A few ideas for modernizing holiday meals.

THE HOLIDAYS ARE HERE—time to set to work in the kitchen. Should we be satisfied with cooking a turkey with medieval methods now that we are living in the twenty-first century? No, let's invent new dishes. But how? The cook who looks to chemistry and physics for inspiration will not find it difficult.

Let's begin by considering a new mode of cooking based on a remark by François Pérégo, a restorer of paintings in Bécherel and a keen student of chemistry who uses egg in treating canvases. He pointed out the effect of ethyl alcohol on egg whites, which you can see by means of a simple experiment: Put an egg white in a bowl and then add a shot of grain alcohol (190 proof). You will discover that the white quickly coagulates.

How can we explain this phenomenon? Albumen is made up of about 90% water and about 10% proteins. These molecules consist of amino acids (twenty types of which are found in foods) that are distinguished by their lateral chains, which may be either hydrophobic or hydrophilic. In water the hydrophilic parts fold up over the hydrophobic parts, minimizing the contact of these latter parts with the water.

The addition of a very strong alcohol alters the environment of the proteins, causing them to unfold. A reaction between two thiol (–SH) groups of neighboring proteins creates a disulfide bridge—a bond between two sulfur atoms—that binds the proteins. Thus a gel is formed, in effect cooking the egg white. Naturally this chemical process cannot be used for culinary purposes without modification: The egg white is tasteless, and alcohol in a nearly pure state holds little appeal from the point of view of flavor. But what if we were to replace the alcohol with a yellow plum brandy, for example, to make a *blanc d'œuf à la mirabelle?* That would be an unusual cocktail to put your guests in the holiday spirit!

Reasoned Braisings

In thinking about a novel main course for our meal it will help to keep in mind some basic scientific facts about the transformations undergone by heated meat. At 40°C (104°F) proteins unfold, becoming denatured, and the meat loses its transparency; at 50°C (122°F) collagen fibers, the chief structural component of muscle cells, contract; at 55°C (131°F) myosin, one of the principal proteins of muscle cells, coagulates and the collagen begins to dissolve; at 66°C (151°F) the sarcoplasmic proteins that make up collagen coagulate; and at 79°C (174°F) actin, another important muscle protein, coagulates.

What use can we make of this information? Instead of cooking your turkey in the usual way, remove the white meat, slice it, and cook some pieces at 50–55°C (122–131°F), others at 55–66°C (131–151°F), and so on (either in the oven, if the temperature control is very precise, or in a pan with stock, using a thermometer). Your guests will be able to enjoy different textures in a single meat because different compounds in the meat will have reacted with one another.

Foamy Foie Gras and Cheese

Now for the cheese course. Earlier we saw how to make Chantilly chocolate, a mousse made of (rather than with) chocolate. Why not repeat this experiment with foie gras? Pour a glass of duck stock into a pan, add some foie gras, and stir over low heat. The melting foie gras forms fatty droplets that are dispersed in the water. Whisk this mixture in a bowl over ice cubes and you will achieve the consistency of whipped cream.

The same thing can be done with cheese. Pour a glass of water or vinegar in a pan, add a sheet of gelatin and a good chunk of cheese (Roquefort, for example), and stir over low heat. The melting cheese forms fatty droplets that are dispersed in the water and covered by the tensioactive molecules of the gelatin. The result is a "cheese béarnaise" that is a cousin not only of many other culinary emulsions, such as mayonnaise, béarnaise sauce, and cheese fondue, but also of cream. Just as chilling and whisking crème fraîche at the same time yields Chantilly cream, beating a cheese béarnaise over a bed of ice cubes will give you Chantilly cheese.

Imaginative cooks are sure to take advantage of this innovation. Didier Clément, chef at the Lion d'Or in Romorantin, has already proposed a goat cheese Chantilly made with Chavignol, accompanied by caramelized shallots. Rather than caramelize the shallots in the usual way with sucrose (ordinary sugar) and a bit of water, Clément was inspired by chemistry to substitute fructose, which produces an original flavor that puts one in mind of candied grapes.

Chantilly Rescued

Let's finish with true Chantilly cream, which will be a perfect accompaniment for your dessert. It is simple to make because one has only to whip very cold cream. Whisking introduces air bubbles into the emulsion, and the fatty droplets crystallize on the outside of the bubbles, stabilizing them. Alas, whisking by hand often is replaced by the electric mixer, which brings with it a greater risk of turning the cream into butter (the result of the fatty droplets fusing and air being lost in the process).

Yet even in the event of mishap all the ingredients of a good Chantilly cream are still present. It is merely a matter of reconstituting them. Simply heat a tablespoon of water in a pan and add the butter. Then put the pan on ice and whisk it once more; the cream comes back. Happy holidays!

100

The Hidden Taste of Wine

Adding enzymes to grape juice releases its flavors.

GRAPES ARE LIKE A LAZY STUDENT who can do better. In addition to odorant volatile compounds (members mainly of the class of terpenols—linalol, geraniol, nerol, citronellol, alpha-terpineol, linalol oxides, and terpenic polyols—whose very low threshold of olfactory perception plays an important role in giving wines their typicity), grapes also contain, in much greater quantities, terpenic glycosides, molecules composed of terpenols bound to sugars. These molecules are precursors of terpenols, but unfortunately they do not contribute to the flavor of wines.

Claude Bayonove and his colleagues at the Institut National de la Recherche Agronomique Laboratoire des Arômes et Substances Naturelles in Montpellier wanted to know whether the aromatic qualities of wines could be intensified by dissociating the two parts of these precursors (sugars and terpenols) by means of acids or enzymes. Because enzymatic hydrolysis seemed more promising than chemical treatment, in part because it gives a more natural aroma, they began by characterizing grape enzymes that release terpenols from their precursors.

Working with glycosidic extracts from grapes of the Muscat of Alexandria variety, the Montpellier team added thirty-four commercially developed enzymatic preparations (pectinases, cellulases, hemicellulases, and so on) to see whether some of them formed terpenols from their precursors. Five proved to be effective, releasing linalol or geraniol depending on the case. All the effective preparations contained beta-glucopyranosidase and alpha-rhamnopyranosidase or alpha-arabinofuranosidase as active components, which were shown to carry out the enzymatic hydrolysis of the terpenic glycosides of the grape in two stages.

This analysis was followed by an in vitro replication of this hydrolysis using rhamnopyranosidase, arabinofuranosidase, and glucopyranosidase. These enzymes released not only the desired odorant terpenols but also norisoprenoids, volatile phenols, and benzylic alcohol, all compounds with very low perception thresholds and an agreeable smell.

The Enzymes of the Grape

The second stage in the glucopyranosidase-mediated hydrolysis of glycosides limits the release of terpenols from both the grape and the wine, for natural enzymes have little effect on the monoglucosides of tertiary alcohols (linalol, terpineol) that have a nonglucosidic part (called aglycone). By contrast, the beta-glucosidase found in yeasts used in winemaking shows weak activity for linalyl-beta-glucoside, one of the principal glucosides of the Muscat of Alexandria grape.

Whereas this grape exhibited noticeable beta-glucosidase activity, it presented only very weak rhamnopyranosidase activity and no activity whatever for arabinofuranosidase, which blocks metabolism in the first stage and limits its overall action on the grape's terpenic glycosides.

Finally, because the grape's glucosidase is unstable and inactive at the level of acidity found in the must and in the wine, it does not seem to be a good candidate for glycoside hydrolysis in either one. Would plant enzymes or microorganisms do a better job of hydrolyzing the terpenic glycosides than the enzymes of the grape? Plant enzymes hydrolyze only the glycosides of primary alcohols, such as geraniol, nerol, and citronellol; the beta-glycosides of tertiary alcohols, such as linalol and alpha-terpineol, were hydrolyzed by only one of the two enzymes studied, though with greater difficulty.

Today the Montpellier physical chemists are studying exogenous enzymatic preparations created by commercial researchers at Gist-Brocades, S.A., from the stock of legally approved microorganisms, seeking to identify organisms that display higher levels of activity at the temperatures and sugar and acid concentrations found in the juice of grapes and in musts.

Professional tasters are sensitive to the presence of intensified flavor in the juice of fruits or of "enzymed" wines, but the use of enzymatic preparations remains largely unexplored. The discovery of new glycosides (such as a recently identified apiosylglycodide) and corresponding enzymatic processes will be of particular importance in improving exogenous preparations.

101

Teleolfaction

Waiting for a new form of telecommunication.

RECALL THE BOOM IN POPULARITY that classical music underwent in the late nineteenth century, when anyone who had access to a telephone could appreciate the virtuosity of the great performers. At the time the transmission of visual images seemed a utopian dream, but only a few years after Alexander Graham Bell patented his famous device Paul Nipkow was awarded a patent in Germany for an apparatus that would transmit such images. This time the beneficiaries were the followers of Polymnia, Terpsichore, Erato, Melpomene, and Thalia.

What realms of communication are left to conquer? The transmission of tactile stimuli is now being mastered, thanks to the design of special gloves fitted with piezoelectric crystals that register or exert pressure. But smells? Flavors? The delay in developing teleolfaction and telegustation is a source of frustration for gourmets.

Olfactory Stimulation

Olfactory and gustatory sensations arise from the binding of odorant and sapid molecules with receptor cells in the nose and mouth. Two possible methods for transmitting such sensations at a distance may be entertained. One could imagine encoding the electrical activity produced in the brain by smell and taste in a series of signals that would stimulate the brain of the receiver by means of electrodes. This is what André Holley and Anne-Marie Mouly at the University of Lyon have been trying to do, hoping to be able to condition rats through excitation of the olfactory bulb.

Alternatively, it might be possible to analyze mixtures of odorant and taste molecules in the same fashion as colors, associating them with arbitrarily selected stimuli. Then, once the electronically encoded information has been transmitted and received, the basic molecules of these stimuli would be combined to reproduce the initial sensations.

Sensors called artificial noses might be useful components for the realization—still a remote prospect—of such systems. Originally developed by the food industry, such sensors are already used in certain processing plants to provide an objective evaluation of the volatile compounds emitted by food. Several types exist.

Artificial Noses

At the Institut National de la Recherche Agronomique (INRA) station in Theix, Jean-Louis Berdagué and his colleagues developed a mass spectrometry system capable of analyzing all volatile compounds in a given sample. Samples are first thermically decomposed in a heated cup. After fragmentation and ionization, the smoke molecules pass into a mass spectrometer, where magnetic and electric fields bend the trajectories of the various molecules in proportion to their mass and electrical charge. Finally, a sensor identifies the quantity of each kind of fragment. In this way the Theix researchers were able to obtain a unique electronic signature for any given mixture. The results are truly wonderful: Such a signature makes it possible to pinpoint the origin of a particular oyster along the coastline of France. What gastronome could match this feat?

Another method involves the use of doped semiconductors on which volatile compounds are reversibly adsorbed, diminishing the electrical resistance of the semiconductor. At the INRA laboratory in Dijon, Patrick Mielle and his colleagues are studying networks of sensors composed of a ceramic substrate, heated by a resistant element, that is coated with a semiconductor such as tin oxide and doped with zinc, iron, nickel, or cobalt oxides. These sensors react differently to the various adsorbed molecules. Whatever the exact mixture of volatile compounds, contact with the sensor network produces a distinctive electrical signature that is then processed by a computer.

The practical application of these networks remains problematic. First, because the signals of the various sensors slowly drift over time, the Dijon researchers devised a rapid transfer system in which compounds are measured and characterized by the sensors and their signatures fed into the network with only a slight delay. Second, the sensors were observed to react quite differently, depending on the temperature. This contingency, which must be controlled for during the measuring process, turns out to be an advantage because measurements made at several temperatures make it possible to obtain different reactions from a single sensor, in effect multiplying the number of sensors in the network.

Finally, because recording the sensors' signals and recognizing the signatures of the various compounds make heavy demands on the system's limited processing power, in practice the number of sensors must be limited to fewer than a dozen. Mielle and his colleagues therefore proposed recording the signals at various instants after the injection of the samples into the measuring cell, proportionally increasing the number of data points generated.

The measuring cells being tested today capture 80% of the gaseous phase molecules, and networks consisting of six sensors that record measurements at four temperatures are able to characterize mixtures of volatile compounds in about ten seconds. The capacities of the human nose are no longer quite as unrivaled as they once were.

Glossary

Some readers may ask why a glossary is needed that defines terms such as *butter, egg white, milk, caramel,* and *chocolate.* Let me respond to this perfectly reasonable question with an anecdote. The Hungarian physicist Leo Szilard used to edit a journal. One day he ran into his friend Hans Bethe, who asked him why he edited a journal. "Because," Szilard replied, "I want God to know the facts." "But don't you believe He already knows the facts?" "Yes, but I want Him to be informed of *my* version of the facts."

In what follows I want to give you my idea of foods, molecules, and the reactions that take place in cooking. But rather than give a comprehensive set of strict definitions—*omnia definitio pericolosa!*—let me share instead my thoughts about a few specific terms.

ACETIC ACID: One of a number of acids present in vinegar. Vinegar also owes its taste to malic acid, formic acid, and tartaric acid, among others. The essential step in the transformation of wine into vinegar involves the oxidation (caused by microorganisms of the species *Mycoderma acetii*) of the wine's ethyl alcohol (or ethanol) into acetic acid.

ACIDITY: A food may be said to be acid, or sour, in the mouth when it causes a sensation analogous to that produced by vinegar or lemon juice. Adding sugar changes this sensation by altering the real acidity of the food (as measured by its ability to dissolve metals, for example). Acidity therefore is better characterized in units of pH, with the aid of small strips of paper and electrodes connected to a measuring device (two items that deserve a place in modern cooking, by the way). Because lentils cook better in a basic than an acid medium, you can adjust the pH of the cooking water with the aid of bicarbonate soda and, once the cooking is done, add vinegar to neutralize the unsavory taste of the bicarbonate.

ACTIN: An important protein in cooking because it is abundant in meats and fish. It acts in concert with myosin in the elongated cells of muscle fibers as an essential agent of muscle contraction.

ALBUMEN: The white of an egg—but if you look at it you will see that it's yellow. Why did our ancestors choose such an ill-suited word? In the *Viandier,* a fourteenth-century French cookbook, the egg white was called "albun," from the Latin *alba,* meaning "white." An egg can have different colors, of course, depending on what the hen eats. In a cookbook from the eighteenth century one reads that in spring, when the hen eats beetles, the egg white is bitter and green.

ALBUMIN: An old word, formerly used in cookbooks to denote what today are called proteins. The term is now reserved for globular proteins such as ovalbumin, the principal protein of albumen. *See also* Egg white.

AMINO ACIDS: The building blocks of proteins. Some twenty kinds are found in cooking.

AMYLOPECTIN: Along with amylose, one of the two molecules that are the principal constituents of starch. Instead of being a mere linear chain of glucose subunits, it has a branching structure, also made of glucose subunits.

AMYLOSE: A glucose polymer. Amylose molecules are composed of long chains of identical subunits of glucose molecules.

AROMA: There us no consensus among specialists concerning the meaning of this term. In the view of some authorities it is the name of the sensation created by volatile molecules released during mastication that rise up through the retronasal fossae into the nose. But others, a majority, take it to describe a part of the overall sensation of flavor. When you drink wine, for example, and perceive a flavor of green pepper, you say that it has a green pepper aroma. This effect is caused by both odorant and taste molecules that stimulate the trigeminal nerve (producing a sensation of freshness, for example). Here I use the term to refer to a distinctive sensation that is a component of the overall flavor of a food or wine.

ASCORBIC ACID: Also known as vitamin C. It is responsible for the antioxidant properties of lemon juice, in which it is present in large quantities (37 milligrams per 100 grams). Instead of using lemon juice to prevent pears, bananas, and apples from browning, why not use ascorbic acid? It's cheaper and much more efficient.

ASPARTAME: A sweetener made by the conjunction of two amino acids, L-aspartic acid and L-phenylalanine methylene. This molecule in itself furnishes proof that there is not just one sweet taste but many. Taste it and see for yourself!

ASTRINGENCY: Characteristic of a molecule that binds with salivary proteins and suppresses their lubricating action, producing the sensation of a dry, tightened mouth. Many polyphenols are astringents because their hydroxyl (–OH) groups bind with such proteins. If you take a sip of a very astringent wine, swirl it around in your mouth for a few seconds, and then spit it out into a glass, you will see precipitates—proteins bound to tannins.

AUTOXIDATION: An essential chemical reaction that causes fats to turn rancid. Because it is self-catalyzing, the reaction takes place quickly. To retard it, protect foods against exposure to oxygen, light (whose ultraviolet rays promote autoxidation), and certain metals.

BASES: Purists are alarmed when these are said to be the opposite of acids, yet it is easier and more helpful to describe them in this way. Pedantry is not always a sign of clarity: Who will understand me, apart from those who already know, if I say that acids release protons whereas bases capture them?

BITTER: This annoyingly vague word is used to denote one of the sensations given by the gustatory papillae. Why annoyingly vague? Because each of us has a very particular impression of tastes and because we do not know which tastes go to make up the flavor of a given dish. In the mouth foods release volatile molecules that travel up to the nose through the retronasal fossae along with molecules that give a sensation of spiciness, while mechanical and thermal sensors in the mouth detect texture and temperature. We therefore have an overall sensation that is customarily called flavor, but no one has ever succeeded in identifying its component tastes in the course of everyday eating and drinking. The problem is compounded by the fact that we now know there is not just one bitter taste but several; electrophysiologists recently demonstrated that the papillary cells that detect quinine (a "bitter" molecule), for example, differ from those that detect denatonium benzoate (another "bitter" molecule).

BOILING: Not the same thing as evaporation. For example, a bowl of water left to sit out for a long time is emptied through the evaporation of the water at room temperature. By contrast, evaporation accompanies boiling when a pan of water is placed over high heat. At what temperature does water simmer? Knowing the answer is important because meat is more tender when it is simmered.

BROWNING: Meats heated at high temperatures brown, as do apples that are cut up and left uncovered, as does sugar that is heated. Browning has various causes, including oxidation, enzymatic reactions caused by polyphenoloxidases, Maillard reactions, and thermal degradation.

BUTTER: A substance that was long considered to be emulsion in which water is dispersed in fatty matter in the form of casein-coated droplets. (Water molecules account for roughly 15–20% of its mass and triglyceride [fat] molecules roughly 80%.) But this view is wrong because at room temperature some of the fat is solid. Butter is better thought of as an emulsion trapped in a solid network, that is, an emulsion contained by a gel.

CALCIUM: This chemical element is found in ionized rather than neutral form in foods, for it has lost two electrons. This divalence makes it an interesting ion, in part because it can bond simultaneously with two pectin molecules, which means that it hardens not only vegetables but also jams. Calcium therefore is neither good nor bad. It is up to the cook either to use it or to seek to eliminate it, keeping in mind that it is readily trapped with citrate ions, for example.

CARAMEL: A delicious brown product that is classically obtained by the thermodegradation of ordinary sugar (sucrose). Different caramels can be obtained depending on the acidity or alkalinity of the medium in which a sugar is heated. Sugars such as glucose and fructose yield distinct flavors from those of sucrose. I love the fructose flavor!

CARAMELIZATION: A term that cooks use indifferently to denote the browning of meats and the thermal degradation of sugars. Yet the chemical reactions are quite different in the two cases. If progress is to be made in the domain of cooking, its terminology must be made more exact: A cat is not a dog, and it would be a poor biologist who confused the two.

CASEINS: Proteins present in milk, where they are assembled into calcium phosphate–cemented aggregates known as micelles.

CELL: Roughly speaking, a small sac filled with water. The wall of the cell is essentially a double layer of phospholipid molecules. A closer look reveals that cells contain a host of interesting molecules, including proteins, sugars, lipids, and desoxyribonucleic acid (DNA), which makes these sacs living entities.

CHEMISTRY: A science of central importance to us as human beings, for it is what allows us to live. In fact, it allows us to live for two reasons: On one hand, our organism functions only by virtue of a coordinated set of chemical reactions; on the other hand, cooking is a form of chemistry that creates the foods that we consume. Chemistry sometimes is defined as the science of the transformations of matter, but this is an exaggeration: Many biological phenomena (the act of chewing, for example) and many of the phenomena of particle physics (the annihilation of a particle and its antiparticle) are transformations of matter (in the first case food is divided up, in the second matter is converted into energy) that do not involve chemistry. It is nonetheless accurate to say that chemistry creates its own subject matter and that it is concerned with the structure of atoms and molecules.

CHLOROPHYLL: A platelet-like molecule, with a tail and a magnesium atom at the center, that contributes to the color of green vegetables. The magnesium atom can be dislodged by a hydrogen ion during cooking in an acidic medium. This replacement of magnesium by hydrogen is accompanied by a change of color: The food turns from green to brown.

CHOCOLATE: Let's not kid ourselves: It's almost impossible for chocolate lovers to stay slim. Chocolate is composed mainly of fat (which accounts for roughly 30 grams of every 100 grams of dark chocolate) and sugar (60 grams). This bare description neglects the fact that chocolate has remarkable organoleptic properties and physical properties that are no less wonderful. For example, its fats melt in the mouth (at 37°C [99°F]) but not in the hand (at 34°C [93°F]). Mineral fats are sometimes added, but very few artisanal or industrial producers show their customers the courtesy of honestly and clearly admitting on their labels the difference between true chocolate and chocolate enriched with fats other than cocoa butter, a situation that seems to bother only me.

CLARIFICATION: A medieval culinary procedure that consists of adding egg white to a stock (wines are also clarified, but here we are concerned chiefly with cooking) and then heating it. The egg white traps the particles that cloud the stock, making it possible to obtain a fairly clear result by means of a final filtering. Rather than waste perfectly good eggs, we can use good laboratory filters instead.

COAGULATION: A transformation whose culinary prototype is the transformation of the egg white, a transparent yellow liquid, into a cooked egg white, an opaque white solid.

COLLAGEN: In meats, a fibrous tissue that sheathes muscle fiber cells. It is composed of proteins that are braided in triple strands, which combine in a way vaguely similar to cellulose fibers in paper. When meat is cooked, the thermal agitation breaks the bonds between collagen proteins, and they pass into solution (the bouillon subsequently gelatinizes as it cools because the proteins recombine with one another).

CONCENTRATION: A term that refers to both a physical phenomenon and a form of measurement. The phenomenon is the grouping together of entities in space or time. Thus, for example, there is a concentration of spectators in a stadium for a sporting event. Obviously molecules can also be concentrated, particularly in herbs; for example, anethol molecules are numerous in fennel. Concentration measures this effect. In cooking, the term often is used in connection with the cooking of a piece of meat. This is an odd way of talking: Juices come out from a roast rather than becoming concentrated inside, odorant and taste molecules are formed as a result of chemical reactions only on the surface, and the temperature inside the roast is lower than that of the oven. What, then, is being concentrated here?

CONDUCTION: In cooking, this phenomenon occurs principally by means of heating: The thermal agitation of molecules on the surface of a food is communicated to neighboring molecules inside. These molecules in their turn disturb molecules still further inside, and so on. This is why the internal temperature of foods progressively rises as they are cooked.

CONVECTION: A phenomenon that accelerates exchanges of heat in a liquid by virtue of differences in density between hot and cold parts. A broth cools rapidly because the liquid at the surface cools from contact with the air, so that its density increases; it then falls to the bottom of the bowl, while the hot liquid rises, then cools, and falls to the bottom. By contrast, a thick soup cools thoroughly only at the surface because its viscosity prevents convection. You can create a remarkable gastronomic sensation with a fragrant liquid (wine or hot chocolate, for example) if you put some of the liquid in a glass at room temperature and then, having heated the rest, gently pour it in the glass. Because of differences in density, stratification occurs between the two liquids. Although it is not visible, you will be able to perceive the effect as you drink.

COOKING: A marvelous activity because it is infinitely variable. The cooking of an egg white is not the same as the preparation of a fish with acids in the Tahitian manner or the same as subjecting a fish to the pressure of several hundreds of thousands of atmospheres. Even if one limits oneself to cooking by heating, cooking is a vast phenomenon. For example, a roast of beef is considered cooked even when it is raw in the center (although in this case the temperature of the meat in the center is actually a little lower than if you were to put the meat out in the sun during the summer). At what point can a food be said to be cooked? Rainwater is the product of prior evaporation. Should we therefore considered it cooked?

CUSTARD: An example of the sort of complex physicochemical systems whose study draws extensively on the work of two Nobel Prize winners, the physicist Pierre-Gilles de Gennes (custard is an example of a suspension–emulsion) and the chemist Jean-Marie Lehn (its proteins combine into supramolecular aggregates).

DECANTING: One of the operations that chemists have learned to master and that cooks could perform more expertly if only they were willing to depart from tradition. But who is prepared to assume responsibility for breaking with the ancients?

DECOCTION: Both very ancient chemistry and cooking—even modern cooking—distinguish between maceration, which involves placing a substance in cold water; decoction, which puts it in boiling water; and infusion, which puts it in boiling water and then quickly removes it. This distinction no longer holds when the solvent is oil, which decomposes before the boiling point is reached.

DEHYDRATION: Water may be drawn out from the tissues of vegetables, meats, and fish with sugar or salt, often in order to make the surface dry and to protect foods against deterioration.

DENATURATION: Proteins are said to be denatured when their form changes, for example by unfolding.

DIFFUSION: An important phenomenon in the kitchen. It is because of the diffusion of light that milk is perceived to be white and because of the diffusion of water molecules that fluids are drawn out from the tissues of salted vegetables and meat.

DOUGH: From the physicochemical point of view, doughs are wonderfully complex systems: solid suspensions in which a solid (starch) is dispersed into another solid (a gluten network formed by kneading).

EGG: A food that consists of a shell (accounting for about 10% of the mass of the egg), white (57%), and yolk (33%).

EGG WHITE: A transparent yellow—not white—liquid containing areas of differing viscosity. If one looks closely one also sees slender white strands and a few bubbles. When heated it solidifies, becoming opaque and white. A first microscopic approximation indicates that it is composed of 90% water and 10% proteins. On closer inspection, it can be seen to contain different proteins, including ovalbumin (58% of the total), ovotransferrin (3%), ovomucoid (1%), ovoglobulin (8%), and lysozyme (3.5%). Coagulation begins at 62°C (144°F); at this temperature the ovotransferrin is denatured, forming a delicate, white, almost transparent network that barely retains the liquid trapped inside it.

EGG YOLK: A remarkable substance composed of approximately 50% water and 15% proteins, with a great many lipids. It has a powerful taste and an unusual texture that changes at temperatures above 68°C (154°F).

EMULSION: A dispersion of droplets of one liquid in another liquid that does not mix with it. Mayonnaise is an emulsion of oil in water.

ENHANCERS: Certain preparations are said to be flavor enhancers, but physiological analysis does not yet justify use of this term.

ENZYMES: Proteins that are responsible for the reactions of other molecules.

ETHANOL: An important molecule because it is found in concentrations exceeding 10% in wines and contributes significantly to their gastronomic interest. The discovery of distillation, which allowed higher concentrations of ethanol to be achieved in the form of *eaux-de-vie* or brandies, was a major event in human history. Certain animals apart from human beings have a sort of language, and some have a form of laughter, but none of them knows how to distill.

EVAPORATION: Not the same thing as boiling, as I remarked earlier.

EXPANSION: There is an old idea that boiled meat cooks by expansion. But boiled meat does not expand. On the contrary, it contracts, and it is this contraction that causes its juices to be expelled into the surrounding liquid.

EXTRACTION: Cooking extracts aromatic and taste molecules from various foods. Nonetheless the flavor of dishes sometimes is wrongly attributed to extraction. For example, a stock owes its flavor to chemical reactions between the molecules extracted from the meat; contrary to what is often said, however, it is not an example of cooking by extraction.

FLAVOR: A term that describes the synthetic sensation produced by eating and drinking (including odor, taste, texture, heat, mechanical properties, and so on) that corresponds to the French *goût*. It is a pity that the English word has been imported into French (*flaveur*) and that *goût* usually is translated in English as "taste"! But let us persevere in our campaign against error: The world of tomorrow will be the one that we create today. *See also* Gustation, Sapiction.

FOAM: A dispersion of air bubbles in a liquid or a solid. A stiffly whipped egg white is a foam. A soufflé is also a foam, but the liquid phase is a suspension–emulsion. The sweetened dessert known as a mousse is an example of an uncooked foam.

FRUCTOSE: A sugar found in certain fruits and in the aisles of your local supermarket. It has a remarkable taste.

GEL: A liquid immobilized by molecules that are linked together to form a network. A cooked egg white, for example, is a chemical gel, for the network formed by the proteins is permanent. By contrast, jams and preserves form a physical—that is, reversible—gel.

GELATIN: Proteins extracted from meat that are reassembled in the form of sheets or powder. In hot water these molecules are dispersed; when the liquid cools, their extremities are linked in threes as segments of a triple helical strand, forming a network known as a gel in physics and, in cooking, as a gelatinous stock or jelly.

GLUCIDES: Molecules formerly considered carbohydrates because the first ones to be discovered had the general structure $C_n(H_2O)_n$, containing as much water as carbon. Nonetheless, it is a mistake to believe that these molecules are composed of water molecules attached to carbon atoms. The various atoms of glucides are arranged in a different manner, with the result that the molecules they compose have numerous hydroxyl (–OH) chemical groups that determine their capacities for combination. It would not be imprecise to call these molecules sugars.

GLUCOSE: A very simple sugar that is found particularly in our bloodstream. The blood carries it to the body's cells, which use it as fuel.

GUSTATION: Flavor (*goût*) is the sensation experienced when one eats and drinks—a synthetic, global sensation produced not only by taste and olfactory and visual perception but also by the perception of textures and various trigeminal stimuli. Nor should the papillae be called gustatory, because they communicate only the small part of the overall sensation known as taste. I propose instead to call the papillae sapictive because they detect tastes (*saveurs*). *See also* Sapiction.

HEAT: A form of energy that is characterized in terms of temperature. Strictly speaking, expressions such as "Heat propagates toward the center of a roast" are mistaken. It is more accurate to say that, in each part of a roasted piece of meat, the temperature increases during the course of cooking or that the rate of molecular agitation of the molecules increases in the meat during roasting, with greater average agitation on the periphery.

INFRARED RAYS: Invisible radiation that is abundantly emitted by hot bodies. Infrared rays are detectable with the aid of a thermometer in the spectrum of sunlight dispersed by a prism, where they come after red. You can perceive them by putting your hand next to your cheek without touching it: The warm sensation you feel is created by infrared rays emitted by your hand.

INFUSION: Tea is an infusion. *See also* Decoction.

IODINE: In alcohol solution a very useful substance for detecting the presence of starch: After a few seconds starch granules turn blue (rather than brown).

KURTI: The Oxford physicist Nicholas Kurti (1908–1998), known for his discovery of nuclear adiabatic demagnetization. In 1988 Kurti and I gave the name "molecular and physical gastronomy" to the scientific discipline that studies the chemical, physical, and biological transformations produced by cooking and eating foods. After his death I shortened it to "molecular gastronomy" and gave Nicholas's own name to the workshops devoted to this subject that are held every two or three years in Sicily.

LACTIC ACID: Produced by lactic bacteria from the lactose found in milk. The characteristic sourness of foods such as sauerkraut is caused by lactic acid.

LACTOSE: A milk sugar. Milk sometimes is said to be sweet on account of its presence, but if you pay close attention you will find it has a salty taste.

LIPIDS: Dictionaries of biochemistry make a point of defining them because several types of related molecules are grouped under this term. Many alimentary lipids are triglycerides or phospholipids.

LUMP: A disagreeable structure, the prototype of which is obtained by putting flour in hot water. The starched periphery of the resulting agglomerations prevents the diffusion of water toward the center, which remains dry.

MACERATION: *See* Decoction.

MAGNETIC RESONANCE IMAGING (MRI): Common name for an imaging process that exploits the phenomenon of nuclear magnetic resonance. In hospitals, the term is a euphemism: In deference to public fears of imagined radioactive dangers associated with nuclear magnetic resonance, one speaks instead of magnetic resonance imaging.

MAILLARD: French biochemist Louis-Camille Maillard (1878–1936). Maillard studied medicine and chemistry at Nancy, writing his medical thesis on urinary indoxyl; his thesis in chemistry, on the action of glycerine and sugars on alpha-amino acids, earned him an international reputation. After volunteering for service in World War I, Maillard joined the medical faculty of the University of Algiers as professor of biological chemistry and toxicology, teaching there until his sudden death in Paris. The reaction that bears his name was announced in a three-page paper published by the French Academy of Sciences in 1912.

MAILLARD REACTION: A transformation that begins with the reaction of a sugar and an amino acid. What follows is very complicated, however, and a complete description would fill several volumes. It suffices for our purposes here simply to say that once an Amadori or a Heyns rearrangement has taken place (depending on the nature of the reactive sugar), several parallel paths lead to the formation of brown compounds, notably the ones found on the surface of meats that are cooked at high temperatures.

MEAT: Roughly speaking, a packet of elongated sacs, or muscle fibers. These sacs contain water and proteins, as in the case of egg whites. The muscle fibers are supported by a tissue, collagen, that is made of elongated (rather than globular) proteins and that dissolves when heated in water.

MEMBRANE: The lining of living cells, consisting of a double layer of phospholipids in which various molecules are dispersed, notably proteins and sugars.

MICROWAVES: Electromagnetic waves that were first harnessed by radar. Engineers observed that pigeons that flew in front of the antenna were cooked, which led to the development of microwave ovens. These utensils are the only ones used in cooking today whose principles were not understood in the Middle Ages.

MILK: Mainly water, but it also contains fats dispersed in the form of droplets that are too small to be seen individually by the naked eye, and proteins, also microscopic in size, that are aggregated in micelles.

MONOSODIUM GLUTAMATE: A compound used by Asian cooks because it produces a remarkable taste called umami.

MUSCLE FIBERS: The slender elongated cells (up to 20 centimeters long) found in meats.

MYOFIBRILS: Cellular complexes of actin and myosin that are responsible for muscle contraction.

MYOGLOBIN: A muscle protein that contains an iron atom, just as chlorophyll contains a magnesium atom.

MYOSIN: Along with actin, an essential protein for muscle contraction.

NEURONS: The brain contains nerve cells, or neurons, that transmit and receive signals in the form of neurotransmitter molecules. Once a certain excitation threshold is exceeded, an electrical impulse is propagated through a prolongation of the cell, known as an axon, triggering the release of neurotransmitters, which traverse a synaptic cleft and activate other neurons.

NUCLEAR MAGNETIC RESONANCE (NMR): A nondestructive method for determining molecular composition that exploits the magnetic properties of certain atomic nuclei along with radio waves. No radioactivity is involved. *See also* Magnetic resonance imaging.

ODOR: A sensation that occurs in cooking when one puts one's nose above a pan on the stove: The heated foods release volatile molecules that attach themselves to the receptors of olfactory cells in the nose—hence the invention of the cover, which retains heat and traps the odorant molecules inside.

ODORANT COMPOUNDS: Molecules that are sufficiently volatile to reach receptor proteins, located on the surface of olfactory cells in the nose. Volatility by itself is not sufficient: A molecule is aromatic only if it binds to one of these receptors, thus triggering the excitation of the olfactory cell, which signals to the brain that an aroma (or an odor) has been detected.

OIL: In cooking, a liquid whose molecules are almost exclusively triglycerides.

OLFACTION: The faculty that permits us to detect the presence of molecules in the air above a dish that have evaporated from it and that contribute to the flavor of the food.

OSMAZOME: A mistaken notion that arose toward the end of the eighteenth century. Alcohol was believed to extract a well-defined principle peculiar to meats—the osmazome—that was responsible for their flavor. Why was this idea mistaken? Because the alcoholic extract in question is composed of many sorts of molecules and because the principle changes depending on the meat.

OSMOSIS: A remarkable physicochemical phenomenon that used to be demonstrated with the aid of a pig's bladder immersed in a volume of water. A bladder partly filled with sweetened water was observed to swell up: The sugar is unable to escape the bladder, and more water seeps in from outside. The underlying reason for this phenomenon is the tendency of concentrations of various molecules on either side of a permeable barrier to be equalized.

OXIDATION: An important reaction that has been insufficiently explored in cooking. In recent years it has often been said that the browning of the surface of meats is caused by a chemical transformation known as the Maillard reaction. It is more accurate to say that the Maillard reaction contributes to browning together with other reactions, among them the Strecker degradation and various oxidizing reactions.

PAPILLAE: Drink a glass of milk, stick out your tongue in front of a mirror, and you will be able to make out small round projections on the tongue. These papillae (usually called gustatory, although I prefer to say sapictive because they detect tastes rather than flavor) are composed of cells whose surface supports proteins known as receptors. When a taste molecule interacts with these receptors, the cells are electrically activated and send a signal to the brain indicating that they have detected this particular molecule. The papillae tell us when we ought to stop eating.

PECTIN: Chemists consider it a D-galacturonic acid polymer, a complex definition that can be simplified by observing, first of all, that it is a sugar. The properties of this molecule result from a long chain of atoms consisting of hydroxyl ($-OH$) groups, a methyl ($-CH_3$) group, carboxylic acid ($-COOH$) groups, and methyl ester ($-COOCH_3$) groups. The carboxylic acid groups are important in cooking because they electrically repel one another in a basic medium, with the result that pectins are unable to combine with one another. Therefore in making preserves (which is to say gels created by the association of pectins) the acid groups must be neutralized by H^+ ions. In other words, the cooking medium must be sufficiently acidic.

PECTINASES: Enzymes that degrade pectin. If you want to make apple juice without wearing yourself out, simply add pectinases to an apple and let them act for a while at room temperature.

PEPTIDES: Molecules formed by the bonding of certain amino acids.

pH: A measure of the acidity or alkalinity of a given environment. The scale runs from 0 to 14. Values between 0 and 7 correspond to acid environments and values between 7 and 14 to alkaline (basic) environments. It is inexcusable that pH paper is not commonly found in kitchens today. How else can one determine a solution's relative acidity?

PHENOLICS. *See* Polyphenols.

PHOSPHOLIPIDS: We could not live without these molecules because, along with glycerolipids, they constitute the double molecular layers that form the membranes of living cells. They have a lipid part and a phosphate part (a phosphorus atom surrounded by oxygen atoms), but most owe their distinctive properties to the presence of an electrically charged part and a hydrocarbon part (composed solely of carbon and hydrogen atoms).

PHYSICS: One of the pillars of gastronomy, involving not particle physics or astrophysics but the physics of "soft" or "condensed" matter. Emulsions, foams, and gels are best studied collaboratively by physicists and chemists.

POLYPHENOL OXIDASES: Enzymes that oxidize molecules of the polyphenol class, forming quinones that react to produce brown compounds. These enzymes are responsible for the discoloration one sees in apples that have been cut up and left out, exposed to the air.

POLYPHENOLS: Molecules containing at least one benzene ring (a hexagonal ring with six carbon atoms attached to hydrogen atoms) and hydroxyl (–OH) groups. Tannins are polyphenols, as are many of the molecules that give foods their color.

POLYSACCHARIDES: Another name for complex sugars.

POTATO: A vegetable whose cells have the peculiar property of containing small granules of starch, which absorb water and swell during cooking.

PRECIPITATION: The French humorist Alphonse Allais used to say that water is a dangerous liquid because a single drop is enough to cloud the purest absinthe. This cloudiness is caused by the precipitation of anethol, a component of absinthe. Precipitation is a phenomenon that has been studied by chemists for centuries and that could be put to better use by cooks.

PRESERVES: Pectin molecules in fruits are joined together by cooking, forming a network—or gel—that traps water, sugar, and the various molecules that give fruits their good taste.

PROTEASES (including proteinases): Enzymes that degrade proteins. Fresh pineapple, for example, contains the protease bromelin; papaya, papain; figs, ficin. These enzymes are our enemies when we try to make pineapple, papaya, and fig jellies.

PROTEINS: Chains of amino acids that are longer than peptides.

RANCIDITY: The result of leaving fat exposed to air. *See* Autoxidation.

RECEPTORS: Proteins on the surface of cells that react by means of weak forces with compounds in the cellular environment and trigger various physiological reactions: detection of an odor, a taste, and so on.

RENNET: An extract from the abomasum of calves that permits the formation of certain cheeses made from milk: Casein micelles, once chemically modified by rennet, cease to repel one another and combine to form a gel that traps fats.

RETRO-OLFACTION: As food is chewed in the mouth it is very slightly heated and releases volatile molecules that rise up through the retronasal fossae at the rear of the mouth and reach the nose, where they are detected by olfactory receptors.

SALT: The kind you find in the kitchen is sodium chloride. As one would expect, its taste is usually salty, but in its unrefined state it may contain other salts that give it a bitter taste.

SALTY: The sensation produced by salt and certain other compounds.

SAPICTION: Both French and English ought to adopt this word, which is much more precise than *gustation*. If the physiology of flavor (not taste) is to make further progress we must take into account what we have learned so far. Just as we have abandoned erroneous notions such as phlogiston and caloric, should we not adopt other more useful ones in their place? *See also* Taste molecules.

SAUCES: Remarkable physicochemical systems that accompany a great variety of dishes. Traditionally they are softer than the meats and vegetables they accompany, but their viscosity must be greater than that of water. Mastery of the rheological behavior of sauces is one of the great challenges of cooking.

SODIUM BICARBONATE: Also known as bicarbonate of soda, or baking soda, it is a valuable base in cooking, used to soften lentils and to accelerate the cooking of vegetables.

SOUFFLÉ: A foam that expands not because its air bubbles are dilated by the heat of the oven but because a portion of its water evaporates.

SOUR. *See* Acidity.

STARCH: Dough is made by mixing flour with water. If the dough is then kneaded under a thin stream of water, a white powder composed of minute granules appears. This material, starch, is also found in the cells of potatoes and other vegetables.

STARCHES: Foods that contain starch. They can be identified by a test familiar to children: Pour a little tincture of iodine on them and see whether a purple stain appears that then turns blue.

STARCHING: A process by which starch granules heated in water lose some of their amylose molecules and swell up, forming a starch paste.

STRECKER REACTION: Amino acids react when carbonyl compounds (which have a $-C=O$ group) are present, producing a degradation named after the German chemist Adolph Strecker (1822–1871). This reaction often occurs in cooking because Maillard reactions are sources of carbonyl compounds.

SUCROSE: Ordinary sugar, composed of a glucose molecule joined together with a fructose molecule (both molecules lose their identity during bonding, however). Sucrose is the prototype sugar molecule, but there are many others.

SUGARS: Some are made up of small molecules, such as glucose, fructose, and lactose. Others are made up of large molecules, such as cellulose and pectin.

SUSPENSION: A physicochemical system obtained by the dispersion of solid particles in a liquid. India ink is an example of a suspension. Custard, in which aggregates of egg proteins are suspended in the water contained in the milk, is another.

SWEET: The taste produced by sucrose and other sugars. The sensation varies, depending on the sugar.

TANNINS: Extracted from various kinds of vegetable matter, including wood, they have the property of combining with proteins and iron. For example, a sheet of (protein) gelatin soaked in strong tea (a solution containing tannins) causes the tea to become cloudy.

TASTE MOLECULES: Certain food molecules dissolve in water and then become attached to receptors on the surface of the papillary cells in the mouth. It is sometimes said that excitation of these cells causes a signal to be sent to the brain that a flavor has been detected. But this terminology is confused, for taste receptors register the sensation of specific tastes. The perception of a food's flavor is the product of a whole set of sensations in addition to the sensation of tastes: perceptions of smell, texture, temperature, and so on. It would be more helpful if the term *sapiction* were used to refer to the detection of a particular taste by sapictive receptors that are carried by sapictive cells in the papillae.

TASTES: It was long thought that only four existed (salt, sweet, sour, bitter), but recent advances in neurophysiology have shown that monosodium glutamate, for example, has a distinctive taste (called umami) and that various bitter molecules stimulate different papillary cells. But consider licorice. Is it salty? No. Is it sweet? No. Sour? No. Bitter? No. Umami? No. Well, then, what is it?

TENSIOACTIVE MOLECULES: Laundry detergents contain molecules that adhere to the surface of greasy stains, enveloping and detaching them from soiled fabrics. The droplets of water-insoluble matter coated with these tensioactive molecules are dispersed in water and then carried off during the rinse cycle. In cooking, molecules of the same type coat droplets of oil introduced into water to form emulsions. See for yourself and try making an emulsion with a drop of liquid detergent, but don't eat it.

TRIGLYCERIDE: An ester of glycerol, attached to three fatty acids. These molecules, which have the shape of a comb with three teeth, make up dietary fats. The teeth—the fatty acids—come in different lengths, and some have carbon atoms that are joined together by double bonds. These unsaturated compounds determine the fusion properties and nutritive properties of foods.

VANILLIN: A surprising molecule that is abundantly present in vanilla. It is also formed when alcohols are left to age in oak casks: The ethyl alcohol reacts with the lignin of the wood, eventually producing vanillin—hence the vanilla flavor of certain old alcohols.

VEGETABLES: Plants whose cells differ from those of animals in that they contain hard walls, softened by cooking. These walls are remarkable structures made up of several layers. It is important for culinary purposes to know that they contain pectin, a complex sugar that causes preserves to set, for example.

YEASTS: Very useful unicellular organisms that help us make bread, wine, beer, and many other dishes.

Further Reading

Abecassis, J., Cakmakli, V., and Feillet, P. (1974). La qualité culinaire des pâtes alimentaires: méthode universelle et objective d'appréciation de la fermeté des pâtes cuites. *Bull. EFM* 264:301–303.

Adenier, H., Chaveron, H., and Ollivon, M. (1993). Mechanism of fat bloom development on chocolate. In G. Charamlambous, ed. *Shelf Life Studies of Food and Beverages*. Amsterdam: Elsevier.

Agrawal, K. R., et al. (1997). Mechanical properties of food properties responsible for resisting food breakdown in the human mouth. *Arch. Oral Biol.* 42:1–9.

Anonymous. (1951). *La cuisine, considérée comme un des beaux arts: livre de chevet de la maîtresse de maison, suivi du Florilège de la cuisine française*. Paris: Tambourinaire.

Anton, M. and Gandemer, G. (1997). Composition, solubility and emulsifying properties of granules and plasm of egg yolk. *J. Food Sci.* 62:484–487.

Arfi, K., et al. (2002). Production of volatile compounds by cheese-ripening yeasts: requirement for a methanethiol donor for S-methyl thioacetate synthesis by *Kluyveromyces lactis*. *Appl. Microbiol. Biotechnol.* 58:503–510.

Atkins, P. W. (1990). *Physical Chemistry*. New York: W. H. Freeman.

Audot, L.-E. (1847). *La cuisinière de la campagne et de la ville, ou nouvelle cuisine économique*. Paris: P. Audot.

Augé, R., Bourgeais, P., and Péron, J. Y. (1989). Étude des conditions de la germination des semences de cerfeuil tubéreux *Chaerophyllum bulbosum* L. *Acta Horticult.* 242:239–248.

Bailey, M. E. and Um, K. W. (1992). Maillard reaction products and lipid oxidation. In A. J. St. Angelo, ed. *Lipid Oxidation in Food*. New York: American Chemical Society.

Balas, L. (1993). *Tanins catéchiques: isolement, hémisynthèse et analyse structurale par RMN 2D homo et hétéronucléaire.* Ph.D. thesis, University of Bordeaux–II.

Beck, I., et al. (1999). Sensory perception is related to the rate of change for volatile concentration in nose during eating of model gels. *Chem. Sens.* 24:275–291.

Belitz, H. D. and Grosch, W. (1989). *Food Chemistry,* 2nd ed. Heidelberg: Springer Verlag.

Berdagué, J.-L., et al. (1991a). Volatile components of dry cured ham. *J. Agric. Food Chem.* 39:1257–1261.

Berdagué, J.-L., et al. (1991b). Volatile compounds of dry cured ham: identification and sensory characterization by sniffing. *Proceedings of the 37th International Congress of Meat Science and Technology.* Kulmbach, Germany.

Bige, L., Ouali, A., and Valin, C. (1985). Purification and characterization of a low molecular weight cysteine proteinase inhibitor from bovine muscle. *Biochim. Biophys. Acta* 843:269–275.

Blanc, R., Kurti, N., and This, H. (1994). *Blanc Mange.* London: BBC Books.

Bonnarme, P., et al. (2001a). L-Methionine degradation potentialities of cheese ripening micro-organisms. *J. Dairy Res.* 68:663–674.

Bonnarme, P., et al. (2001b). Sulfur compounds production by *Geotricum candidum* for L-methionine: importance of the transamination step. *FEMS Microbiol. Lett.* 205:247–252.

Bouzidi el Mehdaoui, S., Gandemer, G., and Laroche, M. (1993). Influence du chauffage sur la composition de filets de truites fario (*Salmo trutta*) élevées en mer. *Sci. Aliments* 13:221–228.

Brady, J. D., Sadler, I. H., and Fry, S. C. (1996). Di-isodityrosine, a novel tetrameric derivative of tyrosine in plant cell wall proteins: a new potential cross-link. *Biochem. J.* 315:323–327.

Breslin, P. A. S. and Beauchamp, G. K. (1995). Suppression of bitterness by sodium: variation among bitter taste stimuli. *Chem. Sens.* 20: 609–623.

Breslin, P. A. S. and Beauchamp, G. K. (1997). Salt enhances flavour by suppressing bitterness. *Nature* 387:563.

Brillat-Savarin, J.-A. (1825). *La physiologie du goût.* Paris: A. Sautelet.

Buchheim, W. and Djemek, P. (1997). Milk and dairy-type emulsions. In S. E. Friberg and K. Larsson, eds. *Food Emulsions,* 2nd ed. New York: Dekker.

Byrne, D. V., et al. (2002). Sensory and chemical investigations on the effect of oven cooking on warmed-over flavour development in chicken meat. *Meat Sci.* 61:127–139.

Caicedo, A. and Roper, S. D. (2001). Taste receptor cells that discriminate between bitter stimuli. *Science* 291:1557–1560.

Castelain, C., et al. (1994). Perceived flavour of food versus distribution of food flavour compounds. *Trends Flav. Res.* 3:47–52.

Causeret, D., Matringe, E., and Lorient, D. (1991). Ionic strength and pH effects on composition and microstructure of yolk granules. *J. Food Sci.* 56:1532–1536.

Chadenier, H. and Chaveron, H. (1995). Comportement physique des mélanges binaires beurre de cacao-matières grasses du lait par résonance magnétique nucléaire pulsée: courbes isosolides et cinétiques de cristallisation. *Oleag. Corps Gras Lip.* 2:237–244.

Chamba, J.-F. (2000). L'emmental, un écosystème complexe. *Sci. Aliments* 20:37–54.

Chevreul, M.-E. (1835). Recherches sur la composition chimique du bouillon de viande. *Journal de Pharmacie* 21:231.

Chong, E. (2001). *L'héritage de la cuisine chinoise.* Paris: Hachette.

Cotterill, O.J. and Stadelman, W.J. (1973). *Egg Science and Technology.* Westport, Conn.: AVI Publishing.

Crank J. (1975). *The Mathematics of Diffusion.* Oxford: Oxford University Press.

Culioli, J. (1994). Le chauffage de la viande: incidence sur la dénaturation des protéines et la texture. *Viandes Prod. Carnés* 15:159–164.

Culioli, J., et al. (1990). Propriétés gélifiantes des protéines myofibrillaires et de la myosine. *Viandes Prod. Carnés* 11:313–314.

Culioli, J., et al. (1991). Propriétés thermogélifiantes de la myosine: influence du degré de purification et du type musculaire. *Actes de la Colloque Science des Aliments, Quimper.*

Cuvelier, M.-E., Berset, C., and Richard, H. (1990). Use of a new test for determining comparative antioxidant activity of BHA, BHT, α- and γ-tocopherols and extracts from rosemary and sage. *Sci. Aliments* 10:797–806.

Cuvelier, M.-E., Berset, C., and Richard, H. (1992). Comparison of antioxidant activity of some acid phenols: structure–activity relationship. *Biosci. Biotech. Biochem.* 56:324–325.

Cuvelier, M.-E., Berset, C., and Richard, H. (1996). Antioxidative activity and phenolic composition of pilot-plant and commercial extracts of sage and rosemary. *J. Am. Oil Chem. Soc.* 73:645–652.

Darriet, P., Lavigne, V., Boidron, J.N., and Dubourdieu, D. (1991). Caractérisation de l'arôme variétal des vins de Sauvignon par couplage CPG-Olfactométrie. *J. Int. Sci. Vigne Vin* 25:167–174.

Darriet, P., Tominaga, T., Lavigne, V., Boidron, J.N., and Dubourdieu, D. (1995). Identification of a powerful aromatic component of *Vitis vinifera var. Sauvignon:* 4-mercapto-4-methylpentan-2-one. *Flav. Frag. J.* 10:385–392.

Defaye, J. and Garcia Fernandez, J.M. (1991). Synthesis of dispirodioxanyl pseudo-oligosaccharides by selective protonic activation of isomeric glycosylfructoses in anhydrous hydrogen fluoride. *Carbohydr. Res.* 251:1–15.

Defaye, J. and Garcia Fernandez, J.M. (1994). Protonic and thermal activation of sucrose and the oligosaccharide composition of caramel. *Carbohydr. Res.* 256:C1–C4.

Defaye, J. and Garcia Fernandez, J.M. (1995). Oligosaccharidic components of caramel. *Zuckerind* 120:1.

De Garine, I. (1993). Food resources and preferences in the Cameroonian forest. In C.M. Hladik et al., eds. *Tropical Forests, People and Food: Biocultural Interactions and Applications to Development.* Paris: UNESCO-Parthenon.

Delaveau, P. (1987). *Les épices. Histoire, description et usage des différents épices, aromates et condiments.* Paris: Albin Michel.

Douillard, R. and Lefebvre, J. (1990). Adsorption of proteins at the gas–liquid interface: models for concentration and pressure isotherms. *J. Coll. Interface Sci.* 139:488–499.

Douillard, R. and Teissié, J. (1991). Surface pressure and fluorescence study of ribulose-1,5-bisphosphate carboxylase/oxygenase adsorption at an air/buffer interface. *J. Coll. Interface Sci.* 143:111–119.

Fauconneau, B. and Laroche, M. (1995). Characteristics of the flesh and quality products of catfishes. *Aquat. Living Resour.* 9:165–179.

Faurion, A. (1987). Physiology of the sweet taste. In D. Otosson, ed. *Progress in Sensory Physiology.* Heidelberg: Springer Verlag.

Faurion, A. (1988). Naissance et obsolescence du concept de quatre qualités en gustation. *J. Agr. Trad. Bot. Appl.* 35:21–40.

Faurion, A. (1993). Why four semantic taste descriptors and why only four? *Proceedings of the 11th International Conference on the Physiology of Food and Fluid Intake.* Oxford.

Feillet, P. (1980). Wheat proteins: evaluation and measurements of wheat quality. In G. E. Inglett and L. Munck, eds. *Cereals for Food and Beverages: Recent Progress in Cereal Chemistry.* New York: Academic Press.

Feillet, P., et al. (1989). The role of low molecular weight glutenin proteins in the determination of cooking quality of pasta products: an overview. *Cereal Chem.* 66:26–30.

Ferhout, H., Bohatier, J., and Guillot, J. (1999). Antifungal activity of selected essential oils, cinnamaldehyde and carvacrol against *Malasseria furfur* and *Candida albicans*. *J. Essent. Oil Res.* 11:119–129.

Fleury, N. and Lahaye, M. (1991). Chemical and physico-chemical characterization of fibres from *Laminaria digitata* (Kormbu breton): a physiological approach. *J. Sci. Food Agric.* 55:389–400.

Franks, F. (1988). *Characterization of Proteins.* Clifton, NJ: Humana.

Gardiner, A. and Wilson, S. (1998). *The Inquisitive Cook.* New York: Henry Holt.

Gardner, R. P. and Austing, L. G. (1962). A chemical engineering treatment of batch grinding. In H. Rumpf, ed. *Zerkleinern Symposion.* Dusseldorf: Verlag Chemie.

Gault, N. F. S. (1985). The relationship between water holding capacity and cooked meat tenderness in some beef muscles as influenced by acidic conditions below the ultimate pH. *Meat Sci.* 15:15–30.

Gay-Lussac, L. J. (1828). *Cours de chimie,* 6th ed. (reissued 1990). Paris: Ellipses.

Gonzalez-Mendez, N., Gros, J.-J., and Poma, J.-P. (1983). Mesure et modélisation des phénomènes de diffusion lors du salage de la viande. *Viandes Prod. Carnés* 4:35–41.

Got, F., et al. (1999). Effects of low-intensity ultrasounds in ageing rate: ultrastructure and some physico-chemical properties of beef. *Meat Sci.* 51:35–42.

Guichard, E. and Pham, T. T. (1994). Le sotolon: marqueur de la typicité d'arôme des vins jaunes du Jura. *Proceedings of the XIIIe Journées Internationales des Huiles Essentielles,* Dignes les Bains.

Guichard, E., Pham, T.T., and Étiévant, P. (1993). Quantitative determination of sotolon in wines by high performance liquid chromatography. *Chromatographia* 37:539–542.

Gunata, Y. (1985). The aroma of grapes. *J. Chromatog.* 331:83–90.

Hagemann, J.W. (1988). Thermal behavior and polymorphism of acylglycerides. In H. Garti and K. Sato, eds. *Crystallization and Polymorphism of Fats and Fatty Acids.* New York: Dekker.

Haslam, E. (1995). Complexation and oxidative transformation of polyphenols. In R. Brouillard, M. Jay, and A. Scalbert, eds. *Polyphenols 94.* Paris: Institut National de Recherche Agronomique.

Hawking, S.W. (1988). *A Brief History of Time.* New York: Bantam.

Hladik, C.M. and Simmen, B. (1996). Taste perception and feeding behavior in nonhuman primates and human populations. *Evol. Anthropol.* 5:161–174.

Huang, Y.T. and Bourne, M.C. (1983). Research note: kinetics of thermal softening of vegetables. *J. Text. Stud.* 14:1–9.

Janovitz-Klapp, M.M., Richard, F.C., Goupy, P.M., and Nicolas, J.-J. (1990). Inhibition studies on apple polyphenol oxidases. *J. Agric. Food Chem.* 38:926–931.

Kamoun, M. and Culioli, J. (1988). Mechanical behaviour of cooked meat under sinusoidal compression. *J. Text. Stud.* 19:117–136.

Kemp, S.E. and Beauchamp, G.K. (1994). Flavor modification by sodium chloride and monosodium glutamate. *J. Food Sci.* 59:682–686.

Kijowski, J. and Mast, M.G. (1993). Tenderization of spent drumsticks by marination in weak organic solutions. *Int. J. Food Sci. Technol.* 28:37–42.

Kurti, N. (1995). Rumford and culinary science. *Proceedings of the Oxford Food Symposium.*

Kurti, N. and Kurti, G., eds. (1988). *But the Crackling Is Superb.* Bristol: Adam Hilger.

Kurti, N. and This, H. (1995). Soufflés, choux pastry puffs, quenelles and popovers. *The Chemical Intelligencer* 1:54–57.

Lahaye, M. (1991). Marine algae as sources of fibres: determination of soluble and insoluble dietary fibres contents in some "sea vegetables." *J. Sci. Food Agric.* 54:587–594.

Lahaye, M. and Jegou, D. (1992). Chemical and physical-chemical characteristics of dietary fibres from *Ulva lactuca* (L.) Thuret and *Enteromorpha compressa* (L.) Grev. *J. Appl. Phycol.* 4:1–6.

Lahaye, M., Michel, C., and Barry, J.L. (1993). Chemical, physicochemical and in vitro fermentation characteristics of dietary fibres from *Palmaria palmata* (L.) Kuntze. *Food Chem.* 47:000–008.

Laroche, M. (1993). Relations entre les caractéristiques avant et après chauffage de la chair de la truite fario (*Salmo trutta*) élevée en mer. *Sci. Aliments* 13:213–219.

Lavigne, F., Bourgaux, C., and Ollivon, M. (1993). Phase transitions of saturated triglycerides. *J. Physique IV, Colloque C8* (supplement to *J. Physique I,* 1989) 3:137–140.

Lavigne, F. and Ollivon, M. (1997). La matière grasse laitière et ses fractions. *Oleag. Corps Gras Lip.* 4:212–217.

Lavoisier, A. L. (1782). Considérations générales sur la dissolution des métaux dans les acides. *Mémoires de l'Académie des Sciences,* 492.

Lavoisier, A. L. (1793). *Traité élémentaire de chimie,* 2nd ed. Paris: Cuchet.

Lefèvre, F. (1997). *Propriétés thermogélifiantes des myofibrilles et texture de la truite.* Ph.D. thesis, University of Auvergne.

Lefèvre, F., Fauconneau, B., Ouali, A., and Culioli, J. (1998). Thermal gelation of brown trout myofibrils: effect of muscle type, heating rate, and protein concentration. *J. Food Sci.* 63:299–304.

Le Meste, M., et al. (1992). Glass transition of bread. *Am. Ass. Cereal Chem.* 37:264–267.

Le Meste, M., Aynié, S., and Colas, B. (1992). Études des propriétés viscoélastiques du pain de mie. *Ind. Ag. Aliment.* 109:862–867.

Le Meste, M., Davidou, S., and Fontanet, I. (1994). Relation entre le croustillant, le mécanisme de fracture et l'état physique des produits céréaliers peu hydratés. *Ind. Ag. Aliment.* 110:11–15.

Le Meste, M. and Simatos, M. D. (1990). La transition vitreuse, incidence en technologie alimentaire. *Ind. Ag. Aliment.* 107:5–11.

Lepetit, J. (1989). Deformation of collagenous, elastin and muscle fibers on raw meat in relation to anisotropy and length ratio. *Meat Sci.* 26:47–66.

Lepetit, J. (1991). Theoretical strain ranges in raw meat. *Meat Sci.* 29:271–283.

Loiseau, B. (1994). *Trucs, astuces et tours de main.* Paris: Hachette.

Loisel, C., et al. (1997a). Fat bloom and chocolate structure studied by mercury porosimetry. *J. Food Sci.* 62:781–788.

Loisel, C., et al. (1997b). Tempering of chocolate in a scraped surface heat exchanger. *J. Food Sci.* 62:773–780.

Lopez, C., et al. (2000). Thermal and structural behavior of milk fat. I. Unstable species of cream. *J. Coll. Interface Sci.* 229:62–71.

Lopez, C., et al. (2002). Crystalline structures formed in cream and anhydrous milk fat at 4°C. *Lait* 82:317–335.

Lucas, P. W. and Luke, D. A. (1983). Methods for analysing the breakdown of food during human mastication. *Arch. Oral Biol.* 28: 813–819.

Lynch, J., et al. (1993). A time-intensity study of the effect of oil mouth coatings on taste perception. *Chem. Sens.* 18:121.

Malencik, D. A., et al. (1996). Dityrosine preparation, isolation and analysis. *Anal. Biochem.* 242:202–213.

Marangoni, A. G. and Lencki, R. W. (1998). Ternary phase behavior of milk fat fractions *J. Agric. Food Chem.* 46: 3879–3884.

Martin, B. (1991). Quantitative determination of solerone and sotolon in flor sherries by two-dimensional capillary GC. *J. High Res. Chromatog.* 14:133–137.

Martin, B., et al. (1992). More clues about sensory impact of sotolon in some flor sherry wines. *J. Agric. Food Chem.* 40:475–478.

Martin, B., Étiévant, P., and Le Quéré, J.-L. (1991). More clues of the occurrence and flavor impact of solerone in wine. *J. Agric. Food Chem.* 39:1501–1503.

Matricon, J. (1992). *Cuisine et molécules.* Paris: Hachette.

Mauritzen, C. M. and Stewart, P. R. (1963). Disulphide–sulphydryl exchange in dough. *Nature* 197:48–49.

Mayer, A. M. (1987). Polyphenol oxidases in plants: recent progress. *Phytochemistry* 26:11–20.

Mayer, A. M. and Harel, E. (1979). Polyphenol oxidases in plants. *Phytochemistry* 18:193–215.

McGee, H. (1990). *The Curious Cook.* San Francisco: North Point Press.

McGee, H. (2004). *On Food and Cooking,* 2nd ed., New York: Scribner's.

McKee, L. H. (1995). Microbial contamination of spices and herbs: a review. *Lebensm.-Wiss. u-Technol.* 28:1–11.

McKemy, D. D., Neuhasser, W. M., and Julius, D. (2002). Identification of a cold receptor reveals a general role for TRP channels in thermosensation. *Nature* 416:52–58.

McNeill, A. R. (1997). News of chews: the optimization of mastication. *Nature* 391:329.

Michon, T., et al. (1999). Wheat prolamine crosslinking through dityrosine formation catalyzed by peroxidases improvement in the modification of a poorly accessible substrate by "indirect catalysis." *Biotechnol. Bioeng.* 63:449–458.

Miles, C. L. and Lawrie, R. A. (1970). Relation between pH and tenderness in cooked muscle. *J. Food Technol.* 5:325–330.

Monnereau, C. and Vignes-Adler, M. (1988). Dynamics of 3D real foam coarsening. *Phys. Rev. Lett.* 80:5228–5231.

Monnereau, C. and Vignes-Adler, M. (1998). Optical tomography of real three-dimensional foams. *J. Coll. Interface Sci.* 202:45–53.

Morris, J. A., Khettry, A., and Seitz, E. W. (1979). Antimicrobial activity of aroma chemicals and essential oils. *J. Am. Oil Chem. Soc.* 56:595–603.

Mottram, D. S., Szauman-Szunski, C., and Dodson, A. (1996). Interaction of thiol and disulfide flavour with food components. *J. Agric. Food Chem.* 44:2349–2351.

Mulder, H. and Walstra, P. (1974). *The Milk Fat Globule: Emulsions Science as Applied to Milk Fat Products and Comparable Foods.* Wageningen, The Netherlands: Pudoc.

Nelson, G., et al. (2002). An amino-acid taste receptor. *Nature* 416:199–202.

Ollivon, M. (1999). Bases physico-chimiques du fractionnement des lipides. *Oleag. Corps Gras Lip.* 6:2–4.

Ouali, A. (1990). Meat tenderization: possible causes and mechanisms. *J. Musc. Foods* 1:109–116.

Ouali, A. (1991). Sensory quality of meat as affected by muscle biochemistry and modern technology. In L. O. Fiems, B. G. Cottyn, and D. I. Demeyer, eds. *Animal Biotechnology and the Quality of Meat Production.* Amsterdam: Elsevier.

Pearson, A.M., Love, J.D., and Shortland, F.B. (1977). Warmed-over flavor in meats poultry and fish. *Adv. Food Res.* 23: 1–74.

Péron, J.Y. (1989). Les potentialités d'élargissement de la gamme des légumes dans la famille des apiacées: l'exemple du cerfeuil tubéreux *Chaerophyllum bulbosum* L. et du *Chervis Sium sisarum* L. *Acta Horticult.* 242:123–134.

Péron, J.Y., Demaure, E., and Hannetel, C. (1989). Les possibilités d'introduction et de développement de solanacées et de cucurbitacées d'origine tropicale en France. *Acta Horticult.* 249:179–186.

Péron, J.Y., Gouget, M., and Declerq, B. (1991). Composition nutritionnelle du crambé maritime (*Crambe maritima* L.). *Sci. Aliments* 11:683–691.

Pham, T.T. and Guichard, E. (1994). Dosage du sotolon au cours de l'élaboration des vins jaunes du Jura. Poster at the XIIIe Journées Internationales des Huiles Essentielles, Digne les Bains.

Prinz, J.F. and Lucas, P.W. (1997). An optimization model for mastication and swallowing in mammals. *Proc. R. Soc. Lond. B* 264:1715–1721.

Rao, M.V., Gault, N.S.F., and Kennedy, S. (1989). Changes in the ultrastructure of beef muscle as influence by acidic conditions below the ultimate pH. *Food Microstruct.* 8:115–124.

Ratsimba, V., et al. (1999). Qualitative and quantitative evaluation of mono- and disaccharides in D-fructose, D-glucose and sucrose caramels by gas–liquid chromatography–mass spectrometry, Di D-fructose dianhydrides as tracers of caramel authenticity. *J. Chromatog. A.* 844:283–293.

Richard, F.C., et al. (1991). Cysteine as an inhibitor of enzymatic browning: isolation and characterization of addition compounds formed during oxidation of phenolics by apple polyphenol oxidase. *J. Agric. Food Chem.* 39:841–847.

Rouby, C., et al., eds. (2002). *Olfaction, Taste, and Cognition.* Cambridge: Cambridge University Press.

Rouet-Mayer, M.A. and Philippon, J. (1986). Inhibition of catechol oxidases from apples by sodium chloride. *Phytochemistry* 25: 2717–2719.

Rousselot-Pailley, D., et al. (1992). Influence des conditions d'abattage et de réfrigération sur la qualité des foies gras d'oie. *INRA Prod. Anim.* 5:167–172.

Rousset-Akrim, S., et al. (1994). Comment distinguer par évaluation sensorielle les foies gras d'oie et de canard? Étude préliminaire. *Sci. Aliments* 14:777–784.

Rousset-Akrim, S., Bayle, M.-C., and Touraille, C. (1995). Influence du mode de préparation sur les caractéristiques sensorielles et le taux de fonte du foie gras d'oie. *Sci. Aliments* 15:151–156.

Seelig, T. (1991). *The Epicurean Laboratory.* New York: W.H. Freeman.

Senée, J., et al. (1998). The endogenous particles of a sparkling wine and their influence on the foaming behaviour. *Food Hydrocoll.* 12:217–226.

Senée, J., Robillard, B., and Vignes-Adler, M. (1999). Foaming of glycoprotein alcoholic solutions. In E. Dickinson and J. M. Rodriguez-Patino, eds. *Food Emulsions and Foam Interfaces, Interactions and Stability*. Cambridge: Royal Society of Chemistry.

Shafer, N. E. and Zare, R. N. (1991). Through a beer glass darkly. *Phys. Today* 10:48–52.

Shewry, P. R., Halford, N. G., and Tatham, A. S. (1992). High molecular weight subunits of wheat glutenins. *J. Cereal Sci.* 15:105–120.

Taylor, A. J. (1996). Volatile flavor release from food during eating. *Crit. Rev. Food Sci. Nutr.* 36:765–784.

Taylor, A. J. and Linforth, R. S. T. (1996). Flavour release in the mouth. *Trends Food Sci. Technol.* 6:444–448.

This, H. (1987). Combat d'ingénieur pour le goût. *Centraliens* 491:19–20.

This, H. (1993). *Les secrets de la casserole*. Paris: Belin.

This, H. (1994). La cuisson: usages, tradition et science. In *La cuisson des aliments* (7e Rencontres Scientifiques et Technologiques des Industries Alimentaires), Agoral.

This, H. (1995a). *La gastronomie moléculaire et physique*. Ph.D. thesis, University of Paris–VI.

This, H. (1995b). *Révélations gastronomiques*. Paris: Belin.

This, H. (1997a). From chocolate béarnaise to "chocolate Chantilly." *The Chemical Intelligencer* 3:52–57.

This, H. (1997b). Pommes soufflées. *La Bonne Cuisine* 136:68–71.

This, H. (1998a). *La casserole des enfants*. Paris: Belin.

This, H. (1998b). A chocolate foam. *The Chemical Intelligencer* 4:27–31.

This, H. (1999). Questions of temperature. *Int. J. Sci. Wine Vin.* 7:98–102.

This, H. (2001a). *Comptes rendus des séminaires de gastronomie moléculaire*, no. 7. (www.sfc.fr).

This, H. (2001b). La révolution de la gastronomie moléculaire. *Chocolat et Confiserie* 381:43–46.

This, H. (2001c). Surfactants in the kitchen: recent advances in molecular gastronomy. In *Actes des Journées Tunisiennes sur les Tensioactifs et Leurs Usages (JTTU)*. Tunis: La Société de Promotion de la Chimie Industrielle et l'Institut National Agronomique de Tunisie.

This, H. (2002a). Bassines en cuivre et confiture. *Sci. Ouest* 191:7.

This, H. (2002b). Cuisine et émulsions. *Revue Générale des Routes (RGRA)* 809:59–65.

This, H. (2002c). Molecular gastronomy. *Angewandte Chemie* (International Edition in English) 41:83–88.

This, H. (2002d). *Traité élémentaire de cuisine*. Paris: Belin.

This, H. (2003). La gastronomie moléculaire. *Sci. Aliments* 23:187–198.

This, H. (2004a). Molecular gastronomy (I). *World Food Ingred.* 3:22–35.

This, H. (2004b) Molecular gastronomy (II): the paradox of culinary innovation. *World Food Ingred.* 4:33–39.

This, H. and Bram, G. (1998). Liebig et la cuisson de la viande: une remise à jour d'idées anciennes. *C. R. Acad. Sci, Paris, Série IIc* 675–680.

This, H. and Kurti, N. (1994). Physics and chemistry in the kitchen. *Scientific American* 270:44–50.

Tilley, K., et al. (2001). Tyrosine cross-links molecular basis of gluten structure and function. *J. Agric. Food Chem.* 49:2627–2632.

Tomas, A., Courthaudon, J. L., Paquet, D., and Lorient, D. (1994). Effect of surfactant on some physico-chemical properties of dairy oil in water emulsions. *Food Hydrocoll.* 8:543–553.

Tomas, A. and Paquet, D. (1994). Effect of fat and protein contents on droplet size and surface protein coverage in dairy emulsions. *J. Dairy Sci.* 77:413–417.

Touraille, C., et al. (1990). Maturation de la viande bovine: évaluation par des méthodes mécaniques sensorielles et par des consommations. *Viandes Prod. Carnés.* 11:291–292.

Vadehra, D.V. and Nath, K.R. (1973). Eggs as a source of protein. *CRC Crit. Rev. Food Technol.* 4:193–309.

Valade, M., Tribaut-Sohier, I., and Panaiotis, F. (1994). Le mythe de la petite cuillère. *Le Vigneron Champenois* 12:28–34.

Vamos-Giggazo, L. (1981). Polyphenoloxidase and peroxidase in fruits and vegetables. *CRC Crit. Rev. Food Sci. Nutr.* 15:49–127.

Van der Bilt, A., et al. (1987). A mathematical description of the comminution of food during mastication. *Arch. Oral Biol.* 32:579–588.

Van der Glas, H. W. and Van der Bilt, A. (1997). Mathematical modelling of food comminution in human mastication. *Comm. Theor. Biol.* 4:237–259.

Varoquaux, P., Offant, P., and Varoquaux, F. (1995). Firmness, seed wholeness and water uptake during the cooking of lentils (*Lens culinaris cv. anicia*) for "sous vide" and catering preparations. *Int. J. Food Sci. Technol.* 30:215–220.

Viarouge, C., et al. (1991). Effects on metabolism and hormonal parameters of monosodium glutamate (umami taste) ingestion in the rat. *Physiol. Behav.* 49:1013–1018.

Vignes-Adler, M. (1996). La mousse de champagne. *Bull. SFP* 106:27–29.

Wal, J.-M., This, H., and Pascal, G. (2000). Aliments issus ou constitués d'OGM: évaluation de leur innocuité. *Med. Leg. Hospital.* 3:111–115.

Walker, J. R. C. and Wilson, E. L. (1964). Studies on the enzymatic browning of apple: inhibition of apple o-diphenol oxidase by phenolic acids. *J. Sci. Food Agric.* 15: 902–904.

Wenham, L. M. and Locker, R. H. (1976). The effect of marinating of beef. *J. Sci. Food Agric.* 27:1079–1084.

Wille, R. L. and Lutton, E. S. (1966). Polymorphism of cocoa butter. *J. Am. Chem Soc.* 43:491.

Wilson, L. G. and Fry, S. C. (1986). Extensin, a major cell wall glycoprotein. *Plant Cell Environ.* 9:239–260.

Index

Molecular Gastronomy

Does a food product or dish especially interest you?
Don't hesitate to contact me about the best way to prepare it:

Hervé This
Collège de France
11 Place Marcelin Berthelot
75005 Paris
France
e-mail: hthis@paris.inra.fr

We will compare our experimental tests.